The Solar Corridor Crop System

The Solar Corridor Crop System

Implementation and Impacts

Edited by

C. LEROY DEICHMAN

Interdisciplinary Corn Production Agronomist,
Deichman Consulting, Shelbyville, MO, United States

ROBERT J. KREMER

School of Natural Resources, University of Missouri,
Columbia, MO, United States

ELSEVIER

ACADEMIC PRESS

An imprint of Elsevier

Academic Press is an imprint of Elsevier
125 London Wall, London EC2Y 5AS, United Kingdom
525 B Street, Suite 1650, San Diego, CA 92101, United States
50 Hampshire Street, 5th Floor, Cambridge, MA 02139, United States
The Boulevard, Langford Lane, Kidlington, Oxford OX5 1GB, United Kingdom

Notices
Knowledge and best practice in this field are constantly changing. As new research and experience broaden our understanding, changes in research methods, professional practices, or medical treatment may become necessary.

Practitioners and researchers must always rely on their own experience and knowledge in evaluating and using any information, methods, compounds, or experiments described herein. In using such information or methods they should be mindful of their own safety and the safety of others, including parties for whom they have a professional responsibility.

To the fullest extent of the law, neither the Publisher nor the authors, contributors, or editors, assume any liability for any injury and/or damage to persons or property as a matter of products liability, negligence or otherwise, or from any use or operation of any methods, products, instructions, or ideas contained in the material herein.

British Library Cataloguing-in-Publication Data
A catalogue record for this book is available from the British Library

Library of Congress Cataloging-in-Publication Data
A catalog record for this book is available from the Library of Congress

ISBN: 978-0-12-814792-4

For Information on all Academic Press publications
visit our website at https://www.elsevier.com/books-and-journals

Publisher: Charlotte Cockle
Acquisition Editor: Nancy Maragioglio
Editorial Project Manager: Hilary Carr
Production Project Manager: R. Vijay Bharath
Cover Designer: Christian J. Bilbow

Typeset by MPS Limited, Chennai, India

Contents

List of contributors

Kerry M. Clark
University of Missouri, Columbia, MO, United States

C. LeRoy Deichman
Deichman Consulting, Shelbyville, MO, United States

Christian Dold
National Laboratory for Agriculture and the Environment, Ames, IA, United States

David Fleisher
USDA-ARS Adaptive Cropping Systems Laboratory, Beltsville, MD, United States

Jerry L. Hatfield
National Laboratory for Agriculture and the Environment, Ames, IA, United States

Mary K. Hendrickson
Division of Applied Social Sciences, University of Missouri, Columbia, MO, United States

Soo-Hyung Kim
University of Washington, Seattle, WA, United States

Robert J. Kremer
University of Missouri, Columbia, MO, United States

Vangimalla R. Reddy
USDA-ARS Adaptive Cropping Systems Laboratory, Beltsville, MD, United States

Dennis Timlin
USDA-ARS Adaptive Cropping Systems Laboratory, Beltsville, MD, United States

Zhuangji Wang
University of Maryland, Plant Sciences and Landscape Architecture, College Park, MD, United States

Foreword

While serving as Provost and Chancellor of the University of Missouri, I was deeply committed to making the University of Missouri a global university and to emphasizing that all disciplines are critical components of a quality university. Therefore I was responsive to the passionate pleas of LeRoy Deichman to explore some of his research that found amazing productivity in a solar corridor crop system. Knowing that I was then chairing the Board of International Food and Agricultural Development (BIFAD), he challenged me to consider whether solar cropping corridor could help feed a global population of 9–10 billion people by 2050, and contribute to soil improvements in Africa. Smallholder farming enterprises play a critical role in food security and poverty alleviation in Sub-Saharan Africa and we agreed that tapping their potential contribution would require innovative scientific applications and nonstandard approaches.

After all, the sun is the world's most sustainable source of cheap energy and maximizing photosynthesis for food production seemed to make good sense. The solar corridor is designed to take advantage of these resources and their productive capacity. Mr. Deichman called my attention to the path-breaking work of his colleague, Robert Kremer, a member of MU's faculty. Dr. Kremer had analyzed the carbon sequestration that was incorporated into solar cropping systems, a positive environmental factor that agricultural production systems must address. Mr. Deichman then invited us to participate in the 2015 annual meeting of the American Society of Agronomy to explore these issues.

After knowing about the national experts exploring these themes from a scientific perspective, I became more intrigued and wondered why these research findings were not more front-and-center in today's debates. As an agricultural economist and university administrator familiar with the world of professional association, I undertook my own inquiries. Professional associations often allow popular trends and funding availability to shape research agendas. In addition, it is not unusual to find conclusions reached in research summaries that might stretch beyond what the research can convincingly support or be fully grounded in the research literature.

For example, with regard to these authors' findings that broader corn row spacing may yield higher output, more than one scientist argued that the fixed nature of today's cultivation and harvesting machinery made wider rows impractical to pursue. Insightfully, a representative of one of the world's most prominent farm implement companies undermined that concern at the ASA meeting by stating a given, "tell us what is needed to produce the food and we will change the machinery to serve that need." Chapter 7, The solar corridor crop system: a pivotal production system for African smallholder farmers, of this book, written by Kerry M. Clark, illustrates the need for such ready willingness to innovate in adapting machinery to intercropping systems of Sub-Saharan Africa. I found it concerning that scientists could limit their vision so easily and, apparently, without fully realizing the risk of leading researchers in a less productive direction.

Fortunately, this book takes us in a more open-minded direction. World food needs and extreme poverty are too critical to overlook potential breakthroughs, whether or not immediate funding is available, or whether or not their success can be guaranteed. However, once found successful, innovative practices will convince decision makers to change directions.

So, I am proud to be asked to write the Foreword for this bold exploration of the potential for solar corridor cropping systems to contribute to global food needs even as it improves carbon sequestration in soils, one of the more complex aspects of climate change. This book grows out of the conviction of the authors that breakthroughs in knowledge come from scientific exploration that asks new questions about known research and commonly accepted practices of modern cropping systems. They are not impeded by popular thinking, and are willing to push the frontiers of accepted trends and conventional thinking of the day. I am eager to support this exploration, as the accumulated supportive science is persuasive, and alternative models deserve scrutiny.

This book speaks to both scientists and farmers willing to reconsider accepted practices of agricultural production. The analysis presented is driven by the authors' passion to provide a tool that may alleviate hunger while improving sustainable production systems. Even though the content is outside my own direct discipline, I want to provide my sense of each chapter's contribution hoping that others will be encouraged to rigorously explore the ideas explored by contributors of this book.

First chapter posits the scientific premise of the book by clearly explaining the complexities of photosynthetic processes and how they can be enhanced through the plant canopy architecture, the interactions among and variations in efficiency of the process due to wind, water, and light, among other factors. It establishes the scientific foundations of "radiation use efficiency" for the remaining chapters. A sense of excitement is already beginning as we realize that crop production is even more unpredictable than we may have originally thought.

The second chapter by coauthor C. LeRoy Deichman goes to the heart of the book's promise. By carefully interpreting agronomic research on solar cropping systems, the experimental processes are deconstructed and examined carefully, revealing some important surprises. Deichman's own research used a careful and intentional "preliminary corn hybrid screening trial (PST) varietal selection process to discover hybrid corn seed varieties that responded most effectively to solar energy." He argues convincingly that responsive varieties are appropriate to their ecosystem, and outlines practical approaches for conducting farm-specific trials. The promise grows more obvious as each chapter unfolds and the organization of the book is revealed more fully.

Third chapter uses the simulation to consider the critical nature of water use for the solar corridor and place-specific implications for the intercropping component that appears to be so important in cropping systems, particularly among smallholder farmers. What becomes obvious is that rigorous research methods can provide needed answers to each intervening variable affecting photosynthesis. A general lesson to guide future inquiry, especially important as changing climate is experienced.

Soil health benefits of the solar corridor are carefully analyzed in Chapter 4, Soil health benefits of the solar corridor crop system, by coauthor, Robert J. Kremer. Intercropping options affect the soil productivity and sustainability and are critical determinants of the ultimate contributions of the solar corridor to improved food production. Particularly, as smallholder producers in Africa confront serious soil deficiencies, the scientific nature of this chapter becomes a critical guide for success.

C. LeRoy Deichman undertakes the formidable task in Chapter 5, "Putting it all together" by demonstrating how research guides the decisions for using a solar cropping system to achieve greater production and sustainability. The research yields insights into machinery use, livestock interactions for grazing and feeding under alternative cropping systems, protecting erodible land, and employing the results in developing countries where small-scale farming is so critical for food security and poverty diminution.

In Chapter 6, The economic and social conditions of the solar corridor cropping system, Mary K. Hendrickson, a social scientist, examines "the economic and social conditions of the solar corridor cropping system" identifying the likely impacts of such a system for communities in the mid-west. Of special interest in her commentary are the ecological impacts and altered risk perceptions as intercropping makes local market access a more realistic potential. A new sense of resilience and "civic agriculture" are possibilities explored.

Chapter 7, The solar corridor crop system: a pivotal production system for African smallholder farmers, presents more elaborate discussion of the smallholder African farming conditions and explores potential changes that could significantly alter and improve food security. The contributor is deeply involved in the smallholder systems in Sub-Saharan Africa, and presents a realistic assessment of the conditions and challenges to smallholder farming improvements and implications for intercropping under solar corridor conditions.

The authors conclude the book in a final chapter that presents their vision of a healthier and more productive food-producing system that serves the human needs of the future and strengthens the sustainability of agricultural production systems. They boldly look ahead to reveal their sense of the significance of their proposals for environmental health, altered food production capacities, and improvements expected in global food and agricultural conditions.

My great interest in this set of ideas is strengthened by the approach taken by these authors. I am proud that so much of this thinking can be attributed to research in Missouri, including that undertaken at the University of Missouri. The credentials of the authors and their credibility among peers encouraged me to "test" the ideas generally in conversations with prominent scientists at research centers and in federal research agencies. Now, I am ready to join the authors in bringing these ideas before professional peers, that noble jury of scientific credibility. May voices of reason and open mindedness take us to the next steps and call for research support to improve our understanding of the many related research needs that become obvious in such explorations. The stakes are high for over 700 million people who go to bed hungry at night.

Dr. Brady Deaton
Provost and Chancellor Emeritus, University of Missouri, Columbia, MO, United States, and Executive Director, Brady and Anne Deaton Institute for University Leadership in International Development, University of Missouri, Columbia, MO, United States

Preface

Many of the books written about crop production have been in the context of modern or conventional systems of continuous monoculture or short rotations using standard planting patterns without provision for off-season crops (i.e., cover crops) or integration of livestock. Thus the vast body of evidence available to today's corn producer offers guidance on supplying all of the currently known plant nutrients that the (corn) plant needs to produce maximum sustainable yields determined by the most limiting factor, as imposed by any given production environment. The point that a limiting factor is reached varies by location, growing season, and specific steps taken by the grower. It is, indeed, an awesome and most challenging task fraught with the unknowns of subsequent critical weather events during the growing season.

This book recognizes and appreciates the basic importance of the above realities, without challenge or further comment, and attempts to address the next most limiting factor as imposed by insufficient reception of photochemical energy by the lower (corn) canopy via the mature chloroplasts throughout the critical stages of reproductive growth. This volume clearly identifies the principles that underlie the solar corridor concept (SCC) within an ecological management approach. Strengths and weaknesses of SCC components including corn hybrid selection, plant populations, and compatible floor crops are assessed. Current gaps in understanding and critical research needs are identified. Possibilities for farmers to demonstrate innovation with a new paradigm for crop production are presented. We also argue that the solar corridor cropping system (SCCS) may greatly reduce the need for expansion of crop production onto marginal land areas by effectively increasing grain yields on currently used acreage.

An overview of basic maize hybrid variety selection, photosynthesis, water-use efficiency, and soil health benefits relative to the SCCS is provided in this book. An effort has also been made to show the potential broader impact of the SCCS on the socioeconomic structure of rural communities and on its application as a means of supporting food systems in developing countries of the world. We also show how the SCCS can be integrated into agricultural production systems as an effective means of providing commodity and biofuel resources, feed grain, and human food.

Livestock integration and soil conservation are summarized in Chapter 5, "Putting it all together", and indicate what our vision is of the future utilization and benefits as well as the needs of the SCCS.

This book rolls out the SCC with specific details supplemented with useful references to apply the SCCS in a site-specific manner to fit the production environment wherever corn is grown. Therefore, this book is intended for interdisciplinary or discipline-specific research and extension agronomists, other capable agricultural scientists and graduate students, and crop producers who possess the necessary tools, background, and positive mental attitude to adapt to innovative and sustainable crop production systems. We sincerely thank all chapters' authors and the publisher for outstanding cooperation and the invaluable contributions to this book. We would be remiss if we did not recognize the many researchers and innovative farmers who have gone before us who provided a body of insightful and state-of-the-art work to which we are indebted for their contributions.

Many of the researchers and innovative farmers who were involved in the development and deployment of the SCCS in the early years of testing are mentioned within the text of this book. Those who have helped with field testing in recent years that was critical in providing our latest agronomic data include Randy Miles (School of Natural Resources and former Director of Sanborn Field, University of Missouri), Tim Reinbott (Division of Plant Sciences and Assistant Director of Agricultural Experiment Stations, University of Missouri), Bill Evans (Department of Horticulture, Mississippi State University), Randall Smoot (former Director of Greenley Research Center, University of Missouri), and Russ Kremer (Wildwood Farms, Frankenstein, Missouri). Harold Reetz (Reetz Agronomics) and Brady Deaton (former Chancellor, University of Missouri) provided valuable consultation and useful comments during the early drafting of the book. One person we are particularly grateful to is Herb Britt, who scanned and prepared many of the figures and tables with attention to detail that enabled us to focus on other aspects of the book. Finally, we thank Nancy Maragioglio (Senior Acquisitions Editor, Elsevier-Academic Press) for originally suggesting that we write the book and encouragement to get the project going; and Hilary Carr (Editorial Project Manager, Elsevier Inc.) for her considerable assistance throughout the writing and editing process.

We have enjoyed working together to coedit the book and write selected chapters and hope that farmers will consider using the crop production system proposed herein to make their farming operations more sustainable while improving soil and environmental health, and restoring vitality to their rural communities.

C. LeRoy Deichman and Robert J. Kremer

Introduction

With an estimated 30 million people dying of hunger every year, according to http://www.theworldcounts.com/counters/global_hunger_statistics, hunger is the world's largest health problem. Crop production systems for producing food for the world's population have resulted in billions of tons of topsoil lost to soil erosion and billions of tons of CO_2, imported from the carbon cycle when dinosaurs roamed earth, foisted into earth's current carbon cycle, which arguably constitute (in the aggregate), earth's largest long-term dilemma.

Earth has a profound need for man to expeditiously address these issues! In addition to the obvious global needs of the human population, the tragedy, expressed by the pain on the face of the child who is displaced from his/her family farm home by the seemingly entrenched consolidation of smaller farms into larger farms, is typically dismissed as a "necessary" part of progress! Many of those who have not summarily written off the small, declining rural communities often wonder why—and then accept the seemingly inevitable! In this seminal book on the Solar Corridor Crop System (SCCS) we disclose how the SCCS uniquely and positively addresses each of the above with meaningful responses that, with your active participation, can "Make A Difference!"

Accordingly—simply stated—the objective of the SCCS is to maximize the productive spatial value of solar radiation on the crop land where it is deployed.

Background

Since the beginning of civilization, "sowers of the seed" have been dealing with the yield-limiting factors (LF) of crops. For example, irrigation from the Nile River was used to effectively eliminate water as an LF. Beans planted with corn (maize) produced symbiotic-N while squash aided pest control for Native Americans, and it is reported that some of these early farmers buried fish under the corn hill to mitigate the LF of plant nutrients. Artificial drainage was used to effectively deal with soil oxygen as an LF in what is now some of the most productive farmland in the world. After all previous LFs are actually removed, the SCCS is our response to removing the next most LF—incident sunlight—deep, in the corn-field canopy!

Since the development of corn hybrids (Shull, 1908) and a subsequently robust fertilizer industry, the trend line of average yield increases shows an undeniably enviable performance record (Nielsen, 2017).

About 42 years later, "the great bulk of maize throughout the US was hybrid" (Crow, 1998) and yet, 111 years later:

1. People are still starving (30 million per year) while the human population continues to increase.
2. People are still addicted to oil (in apparent disregard of the fact that the biosynthesis of atmospheric carbon, from the current carbon cycle as captured by plants, is a significant, viable fuel source to displace fossil-sourced carbon imported from the carbon cycle when dinosaurs roamed the earth)! The current role of biofuels, while mostly limited to improving the octane and oxygenate content of gasoline is, however, reluctantly embraced.
3. Outstanding farmers are still losing their lands.
4. Numbers of farmers are still decreasing.
5. Rural communities have declined, even disappeared.
6. Soil is still eroding.
7. Clean water is still being polluted.
8. CO_2 in the air we breathe is still increasing.
9. Our most finite agricultural production resource—arable land and its richest topsoil—is still disappearing.

It is universally recognized that food, water, and oxygen are basic, indeed vital, components for human life to exist on earth! Production agriculture's universally recognized principle role is to produce foodstuffs that enable us to feed the world. Since the 1970 Clean Air Act, the demands to replace lead (known to impair brain function), then benzene (a known carcinogen), and more recently methyl tert-butyl ether MTBE (octane boosters in gasoline that caused untold harm to humans) and to displace fossil fuel, based on concerns of excessive CO_2 in the air, have added another critical, if not well-understood and subsequently recognized, human need.

These are the monumental challenges that are proactively addressed by this seminal volume.

Accordingly, this book is about the SCCS, which positively and humanly enables the proactive resolution of each of the above critical facts via a 180° paradigm shift from current well-established and highly acclaimed successful practices, as indicated in each of the last 78 years in which the average US corn (earth's most productive food crop) yield increase has been an amazing 1.8 bushels (bu) per acre per year, or more than quadrupling! Logical details of each fact are intended to unfold in a way never imagined before reading this book.

In consideration of above and the unprecedented productive value and sheer magnitude that photosynthesis contributes to crop yield, it seems logical that we should do whatever possible to enable every productive chloroplast-laden leaf to fully access the abundant, low-cost photochemical energy source delivered by solar radiation. When photosynthetically active leaves are incrementally shaded, potential yield is limited. Where all other crop production factors are optimized, incident sunlight on the mature chloroplasts becomes the LF. The basic mechanism of photosynthesis carried out in chloroplasts is detailed in Chapter 1, Photosynthesis in the solar corridor system (Hatfield, 2019). An appropriate width of "solar corridor" between each row of corn or other main crop, or between a set of twin rows, offers maximum response to the current yield-LF of solar radiation and serves as the core and basis of the appropriate site- and hybrid-specific SCCS. This seminal book on the SCCS offers a brief overview of our experience to date in testing the hypothesis and subsequent synergistic development of the concept and our current best efforts in defining and maximizing the sustainable value of the previously described parameters.

The challenge

The hypothesis, held since its genesis and expanded through current developmental stages of the solar corridor concept (SCC), is that a specific repositioning of solar radiation, deep into the corn field and a specific redirection of human capital to more effectively utilize that abundant renewable resource, may significantly increase productivity. This concept has been developed and validated through scientific evaluations and subsequent seminal studies at several locations in the U.S. corn belt above 36° N latitude (Deichman and Murrell, 1977; Deichman and Barrett, 1978; Deichman, 1980, Deichman and Warsaw, 1988, Deichman, 2000, 2002; Deichman et al., 2001). These results, not only resurrect—but achieve—the challenge posed by Jonathan Swift in his 1727 Part II, Chapter 6 proclamation (https://en.m.wikiquote.org/wiki/Gulliver%27s_Travels).

The level of this challenge begins at:
the highly acclaimed (and justly deserved), state of the art, 21st century production baseline and represents the authors' attempt to maximize the gift that we've been given, in Genesis 1:3.
That performance validation data audaciously shows that we can double the corn yield per spatial unit of our most finite crop production resource—arable land! The growing body of evidence further supports the broader hypotheses that the SCCS enhances many soil-health components and subsequent sustainability of not only soil, mankind, and the planet, but the relative viability of the family farm and the integrated rural community. For example, people are starving.
Vital topsoil is eroding or otherwise disappearing.
Farm-sourced waters that people drink is being polluted.
The CO_2 concentration of air that people breathe is increasing.
Fewer farms are considered necessary and the subsequent disastrous consequences to rural communities has become an accepted reality.
A practical challenge to the authors is simply to determine the most effective way to describe what the SCCS enables in a short description. "The properly designed, developed and deployed SCCS enables the production of as much or more corn from half as many rows of corn as is produced from current best management practices (BMP) implemented in state-of-the-art corn production"—is one such short and inadequate attempt.

1. This sea change order of SCCS corn yield results are hybrid-specific and urge an attitudinal adjustment toward all steps of this paradigm shift.
2. Our experience with the SCCS has been consistently successful with twin rows (one hybrid demonstrated a similar degree of success without twin rows).
3. Vital supporting practices of the solar corridor planting arrangement include a paradigm shift in the selection of (1) corn hybrids and (2) rotational crops; and in the management of (3) rotational crops in the SCCS and sets the stage for an additional 180° paradigm shift in its distribution and uses.

All of the above issues are positively addressed by this volume. The seminal research has been conducted by Maize Research Unlimited, Inc. (Deichman et al., 2001), its core intellectual property has been placed in the public domain by the issuance of a US patent (Deichman, 2000), replicated in field research studies in northeast Missouri (see Table 2 in

Nelson, 2014), confirmed in central Missouri (Bradford Farm), Sanborn Field (Kremer and Deichman, 2014a), and a parallel study in Crescent City, IL.

Chronology

Reminiscent of my outside row observations, Dad's row-width trials (1943–45 soybeans), the perceptive counsel of my parents while growing upon the family farm, my academic education and experience at the University of Mo.; in 1966 I (Deichman) witnessed, the light reflection research being conducted by Dr. Johnny Pendleton and his group at the University of Illinois that shed light and potential insight on the magnitude of an LF, typically unrecognized or inadequately addressed by the corn grower and the crop production supply industry. That LF set up by modern corn production practices and the subsequent inefficiency of mature (based on number of days since metabolically functional) chloroplasts, which as juvenile chloroplasts, enjoyed a full dose of the premier photochemical energy source for photosynthesis—incident sunlight; but, as the plants grow, is deprived of incident sunlight. Energized by Pendleton's research, Deichman in 1967 conceived and established a "very unusual" treatment with two high-performing hybrids of contrasting but complimentary morphological characteristics to address this LF on his Iroquois County, Illinois farm (40°37′N latitude) in north–south rows on Bryce–Swygert (fine-silty, mixed, superactive, mesic Aquic Argiudolls/Fine, mixed, active, mesic Aquic Argiudolls) soils with follow-up plans on the more productive Milford–Martinton soils (Fine, mixed, superactive, mesic Typic Endoaquolls—Fine, illitic, mesic Aquic Argiudolls). Treatment results indicated an amazing potential! The next year (1968) Deichman established another "very unusual" but advanced version of the treatment on Nicollet–Webster soils in Minnesota (43°53′N latitude) with a third hybrid selected by Dr. A. Forest Troyer. Likewise, it promised an unprecedented potential and established a new morphological approach to canopy architecture and its resulting effect on microclimate in the corn field.

Other commercial priorities of the time, while deferring the now established morphological approach, gave Deichman an unprecedented opportunity to not only investigate, but react to, the genetic aspects of the interdisciplinary concept on the Nicollet–Clarion–Webster soils of the northern corn belt (Deichman et al., 1969; Deichman and Murrell, 1977; Deichman and Barrett, 1978; Deichman, 1980). Those findings led to the confounding conclusion that high-performing modern corn hybrids not only differed in their quantitative responses to soil fertility variables, but the genetic heritability tracked to several essential plant nutrients at a highly significant level (BLSD .01). Duane Shields of Shields Soil Service suggested that if I returned to Illinois and joined his agronomy staff, he'd buy the fertilizer plant that champion corn grower, Herman Warsaw, used to further accommodate my research objectives! I knew that Dr. L. Fred Welch planned to conduct some new light studies at the nearby University of Illinois campus, and Jerry Ash, beginning in 1983, would provide use of his new Kinze twin row planter. Thus, I returned to Illinois to "complete" the work at sites in Champaign, Iroquois and McClean counties. Herman Warsaw readily agreed with the concept observing that "insufficient sunlight at the base of the corn plant is limiting my yields" and we began to look at the SCCS system on his farm. With above operational and attitudinal developments firmly established, we were ready to begin the evaluation of the convergence of two critical and basic developmental components of the concept: canopy architecture and hybrid-specific (varietal) response.

We presented our preliminary objectives and mission (Deichman and Warsaw, 1988) of the planned new interdisciplinary field studies. We were essentially attempting to "apply" the "basic" light-enhanced research of Pendleton et al. (1967), my former soils professor, Dr. Ellis Graham (Graham et al., 1972), and Ottman and Welch (1989), among many others. Further interdisciplinary support was provided by professors Woodruff (soils), Wayne Decker (meteorology), and renowned corn breeder, Marcus Zuber, along with all professors at my alma mater, University of Missouri; Dr. Dale Moss, University of Minnesota and Dr. W.H. Pierre, Iowa State University, provided personal support and timely advice that I most sincerely appreciated and utilized before the 1977 paper could be presented. While Duane Shields and those listed above had provided the opportunity, Ed Frichtl of Paxton, Illinois, was the first farmer to take an active interest in the work upon my return to Illinois in 1982. This was backgrounded by Dr. W.A. Albrecht, Chair of the University of Missouri Soils Department and owner of a farm where I worked with three generations of the Knox family; Dr. R.L. Wehunt, Chief Agronomist (mentor of my self-directed postgraduate interdisciplinary corn production studies); Bob Wyatt and Milton Schofield, my supervisors with Schofield Soil Service; Dr. Art Gohlke, Ag Research Director, for the parent company of Custom Farm Services (CFS) who sent a complete new line of research equipment and research agronomist (Barrett) collaborator to my Minnesota research sites; Lab Director, Dr. Darst; Plant Pathologist, Dr. Reynolds, and the complete CFS agronomy team. I conducted my first field-scale planter (8-row, 30-inch widths) studies with Jerry Ash's new Kinze twin row planter in 1983. The 7.5-inch twins turned out to be combine-harvest compatible (the 10-inch twin rows of our Birch planter were not combine-harvest compatible and 5.5-inch twin rows in Welch appeared to be too close).

When Fred Welch began his light studies, published in Ottman and Welch (1988) I was "on the scene" like any true "consummate student of corn production" or like a "bumblebee on clover!" Fred's results and his personal challenge to me to "figure out how to apply it," further incentivized me to do so—thus the promise of the concept offered must surely be developed! Accordingly, I restructured my work to "house" (conduct) the field research. Deichmann Maize Research Inc. (DMRI) drafted the preliminary parameters, established and directed the preliminary hybrid screening trials, and developed the study protocol. Deichman Consulting (DC) conducted the initial work. Rick Nelson offered to host a field research site and prompted me to call a meeting of other DC clients, who promptly responded to my call by stepping forward to host the appropriately diverse and requisite number of advanced study sites. Emerson Nafziger of the University of Illinois helped design the initial field layout. Maize Research Unlimited (MRU), led by President Don Frichtl, Bob Lindgren, Nelson, other DC clients and other involved and subsequently interested parties, was formed to fund and conduct the studies with substantial discounts on all plant and soil sample analyses provided by Brookside Laboratories. Subsequent to our promising initial results, MRU commenced to patent the results (Smith, 2000). The USPTO said the patent was too broad and needed to be split up and submitted as multiple applications. MRU withdrew all but the core application, which was approved and issued in 2000. Due, in part, to exhaustion of funds, the other four withdrawn applications are still in place and pending submission. Subsequent work by Kremer et al. (2017) has identified additional carbon sequestration components eligible for patent submission. About 12 years later, beginning in 2000, the now completed studies proposed by Deichman and Warsaw (1988), serve as the basis of the referenced "seminal work" placed into the public domain by the USPTO (Deichman, 2000) and into the scientific literature by Deichman and others as papers presented at Annual or Biannual International Meetings of various scientific societies most capable of mediating effective critique and/or collaboration on subsequent work. Following up the ongoing research, subsequent developmental work has been presented by collaborators in proceedings of professional meetings and independent peer-reviewed articles in scientific publications and peer-review journals. This, the first book on the solar corridor, offers an overview and specific details of previous and current research, and a renewed vision for a more productive future!

Initial observation, development logic and abundant key bioresources

The SCCS is based on a common observation: an input typically considered to be a necessary element of modern high-yield corn production, the uniquely critical analysis of its basic impact on corn yield, and the difficult, disruptive journey of how to demonstrate, apply, and effectively quantify its potential and/or actual production impact and eliminate it as a potential LF.

The observation:

The lower leaves are deprived of incident sunlight after full canopy development. Its apparent initial impact: the "mature" chloroplasts (MC) are deprived of the vital photochemical energy source that "fuels" photosynthesis through grain fill.

The logical but disruptive response:

This is what The SCCS is all about.

It first challenges us and the unprecedented successful conventional wisdom, and then instructs us on the hybrid-specific, subsequently realized synergistic values of photosynthesis and its audaciously increased levels of corn production enabled by the SCCS.

Three uniquely integrated, operational keys to maximizing solar corridor performance:

1. Abundant, sustainable, cost-effective incident sunlight.
2. Abundant, sustainable, cost-effective, (current cycle) atmospheric CO_2.
3. Abundant human attributes of ambition, commitment, dedication, energy, ingenuity, intellect, labor, persistence and skill; driven by a strong, family farm- and community-based positive mental attitude (PMA).

The properly designed, developed, and deployed SCCS produces extra yield by enabling a larger proportion of the chloroplasts, essentially the

MCs, to more efficiently utilize the incident sunlight received in the field, which enables the plant to effectively capture more CO_2 from the abundant and sustainable current carbon cycle above each field; and more effective utilization of all essential available plant nutrients. While recognizing and building on the vast body of published knowledge of sustainable soil health, crop production, plant genetics, and plant nutrition and experienced on-farm plant husbandry, this seminal volume on the SCCS introduces basic details of how incident sunlight—the prime photochemical energy source—and its synergistic multifactor impact—enables unprecedented cultivar and variety-specific increases in crop yield.

Perhaps there are two basic elements we must recall about the prime source of the SCCS yield increase, (solar radiation)—light and heat. Light contributes to physiological yield. Heat contributes to morphological development. In the barest knuckle assessment, light is good from leaf emergence until physiological maturity and heat is good when soils are cold but bad whenever too hot. One of the several problems with increased heat is, in general terms (which we should seldom rely on), that more heat leads to faster plant development and subsequent reduced time for chloroplasts to intercept incident sunlight needed to maximize productivity.

SCCS development logic

The preliminary variety screening study (PST) is the first step of the testing protocol. If we assume an average trend line yield of 176.6 bu from 30,000 plants per acre, the baseline production is a little less than 1/3 lb of #2 corn per plant. This example requires hybrids that produce significantly less than 2/3 pound per plant in the PST to meet our seminal standards for further consideration as an excellent hybrid choice for the solar corridor. See Chapter 2, Variety selection for the solar corridor crop system (SCCS), for the specifics. With hybrids that meet the above PST for the SCC, an astute farmer, student, or practitioner of corn production can see how corn yield may be effectively doubled. By planting the baseline population with a corn compatible crop on rotational or "floor crop" acres, the SCCS produces both the main crop and the rotational crop in the same field instead of having one crop in one field and the rotational crop in another field (each crop occupies exactly half of the field).

For the corn grower who successfully deploys currently recognized and widely adopted BMP, the SCC offers a uniquely different approach that enables further substantial increases in sustainable corn yields and net operating profits per corn acre. We emphasize "sustainable", based on many soil health criteria, including the SCCS's vital capability to potentially reduce soil erosion, soil compaction, postharvest soil nutrient losses, and lodging losses, while increasing carbon sequestration, active carbon, beneficial microbial activity, water infiltration rates, water storage capacity, nutrient holding capacity, and organic matter (Kremer and Deichman, 2014b). The SCCS accommodates two or more, compatible crops simultaneously in alternating strips within the main crop across the field for more effective use of solar radiation, water, CO_2, and nutrients compared with monocultures of either component crop (Kremer and Deichman, 2014a).

The most obvious production strategy of the SCCS is to enable all active chloroplasts to produce photosynthates to their maximum genetic potential. This requires that many current BMPs are provided. Before the synergies offered by the SCCS can be maximized we must use the unique solar corridor criteria (or equivalent) for selecting optimum performing corn varieties.

The microclimate-altering canopy architecture embodied in the properly designed, developed, and deployed SCCS is delivered by planting a wide-row spacing of tall-stature crops (i.e., corn) to form the solar corridor and a more uniform, vertical distribution of incident sunlight available to all fully exposed leaves from top to bottom of the plant. This forms the solar corridor planting arrangement that accommodates a secondary or cover crop to intercept sunlight reaching open space between the wide rows of corn (Timlin et al., 2014, 2019). Once all site-specific, soil-sourced essential plant nutrients are made available for timely plant uptake using current BMPs, and effectively eliminating plant nutrient availability as an LF for optimum grain yield, incident sunlight and water then become the most LFs. With appropriate irrigation and/or good rain-fed growing seasons, incident sunlight is the biggest LF. The SCCS is not a typical intercropping practice where the only objectives are increased yield and profit of two or more crops per land area unit or curbing food security challenges due to limited arable land and numerous environmental stresses (Liebman and Staver, 2001). The SCCS is designed to maximize efficiency of chloroplast function in both light and CO_2 capture for efficient photosynthesis by each of the selected crops (Kremer and Deichman, 2014a). However, similar to intercropping systems, overall productivity in the SCCS may be more resource-use-efficient and productive than in monoculture and may be particularly adaptable in low-yielding environments under stress (i.e., drought, disease) and inadequate crop nutrition (Temesgen et al., 2017). Thus, the SCCS can be adapted to any production system including conventional commodity crop production on large acreages, ecological or sustainable cropping, organic farming, or subsistence agriculture on small land holdings.

Achievement of increased corn yields then is made possible through:

1. More efficient utilization of photosynthetically active incident sunlight, that is, involvement of every leaf capable of active photosynthesis, and the subsequent increase in the synergistic biocapture of atmospheric-sourced carbon dioxide and more efficient use of all essential nutrients and water. In other words, the properly designed, developed, and deployed SCCS enables higher corn yields by more efficient capture of photochemical energy that drives photosynthesis.
2. The properly deployed skills of the ambitious [capable], committed [competent], dedicated, energetic, innovative, intuitive, and skilled corn grower who is willing to utilize the SCC, with a uniquely PMA, to move beyond the inherent yield limitations of currently accepted BMPs.

Regardless of the design chosen (the proper design depends on the site-specific conditions and objectives of the grower), it enables the corn plant to more efficiently use crop production inputs including water, nitrogen, and other essential nutrients. The other crops in the crop rotation are additive and integral components of that design. Details on devising a properly designed SCCS are provided as guidelines for farmers in later chapters in this book. The additional qualities of the SCCS offer long-term benefits to the sustainability and health of our soil (Kremer and Deichman, 2014b) and to our farms and society.

The SCCS requires a crop production paradigm shift based upon (and in recognition of) the fundamental fact that good arable farmland is the most finite production resource—and not the number of people potentially capable of farming it! In order for society to be sustained in a most humane manner for the future of mankind, productivity per spatial unit needs to be prioritized higher than spatial productivity per people unit. The crop production paradigm, still used today, was initially adopted when the availabilities of those two production resources were reversed. We then had seemingly unlimited productive and arable farmland but a small population potentially capable of sustainably farming all of the land in crop production.

Solar corridor production objectives

The overall objective of the SCC is to increase sustainable productivity and net operating profits per spatial land unit.

The name of the concept comes from the fact that solar radiation in the form of incident sunlight is the basic factor that enables corn producers to increase production. A specific objective of the SCCS is to maximize the effective reception of incident sunlight and its contribution to photosynthetically active radiation (PAR) and radiation use efficiency (RUE). The premier photochemical energy source and atmospheric CO_2, trapped by the plant, are basic to crop productivity. Remarkable increases in the state of the art, BMP corn production have been largely due to successful manipulation of previously defined LFs using modern management thereby leaving effective access to incident sunlight and CO_2—as LFs in modern corn fields. Thus, the SCCS is a means to move beyond the inherent limitations of currently accepted BMPs. For illustration purposes, let's assume the SCCS is applied as a "whole farm" paradigm shift (WFPS) in a two-year rotation of corn and winter wheat/stubble cover crop (Fig. 1). The WFPS changes the current conventional monoculture system by rearranging the rotation of two individual fields every other

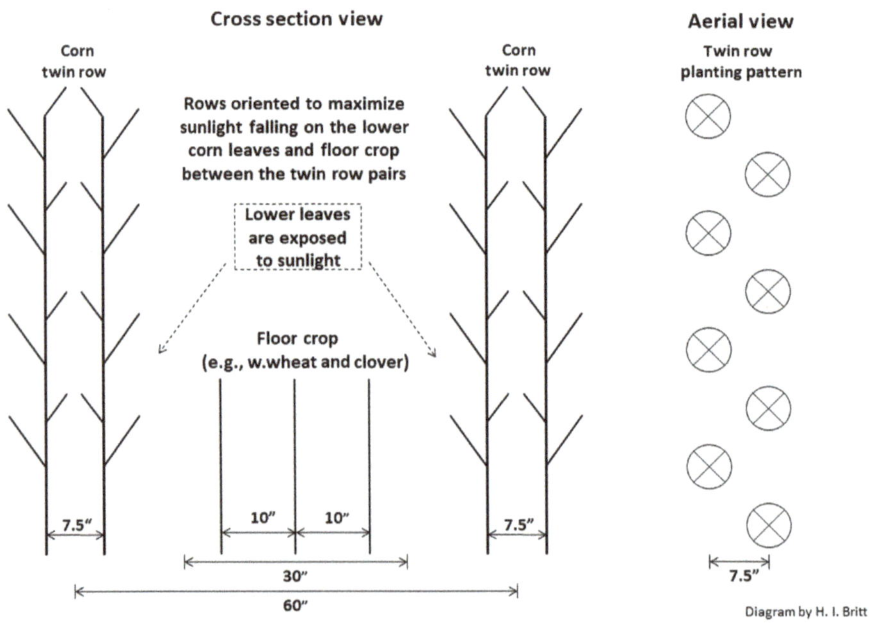

Figure 1 Schematic of the solar corridor cropping system.

year between two crops over two years into one large field. The conventional system calls for two fields that are rotated between corn in year 1 and a winter cereal grain crop/stubble cover crop in year 2. Implementing the WFPS uses the same area but in one large field planted to both crops in each year. The grain/stubble crop is designated as the floor crop, on the other half of the field, within the corridor between the wide-row planted corn. Once the corn begins to canopy and deprive the winter cereal crop of incident sunlight, the cereal crop has already matured physiologically and is ready to be harvested as forage or grain. Subsequent research in field trials conducted by Nelson (2014) in a simulated WFPS revealed a doubling of spatial grain yield for one corn hybrid planted in a SCCS arrangement (see Table 2; Nelson, 2014).

The original SCCS mandate and design (Deichman, 2007) called for floor (corridor) crop species that did not interfere with, or reduce, agronomic productivity of corn yields (Deichman, 2000, 2009). Preliminary observations of a stubble crop established in small cereal grain harvested from the corridor suggests a much higher stubble forage yield using the new paradigm compared with conventional rotations. It is also reasonable to believe cereal grain yields would be as much or higher using the new paradigm. Yields of another crop (e.g., *Little Marvel* peas) support the claim that a "floor crop" can be produced without reducing corn yield (Deichman et al., 2016). Much work is required to determine the optimum species and varieties of the second year, or rotational floor, crop in this system, shown in Fig. 1. The authors have observed many floor crops that perform well in this system. The corn emphasis is not intended to limit main crop species to corn. Notably, preliminary observations show that soybeans respond to the SCCS and alternative C-4 crops such as grain sorghum are adaptable to the SCCS. Considerable effort remains in assessing the many options and opportunities that the SCCS will enable. We are confident that if those knowledgeable in the field keep the original mandate and adopt the changes, modifications, or revisions to fit specific needs, benefits will become apparent as they unfold the concept and achieve success in sustainable crop production.

Potential impacts

"We think we know" that corn growers around the globe can double their corn yield per spatial unit area. We say "think" rather than "know" because we have not conducted evaluations on a global basis and we personally no longer (as we had in Deichman and Barrett, 1978; Deichman, 1980) have knowledge of the specific genetics of the commercial corn hybrid, let alone its quantitative response to the SCCS, sufficient to select proprietary hybrids that predictably respond to the SCCS in terms of productivity. Hybrid selection for the solar corridor presented in Chapter 2, Variety selection for the solar corridor crop system (SCCS) (Deichman, 2019) will clarify our current status on this critically important issue by selection of the proper hybrids for the SCCS.

We can increase the fixation of CO_2 to achieve:

1. higher yields throughout the double peaks of grain fill;
2. over longer periods of time; and
3. more fixed carbon, the major source of organic matter, sequestered into the soil.

Let us set forth to demonstrate how your SCCS can achieve these objectives!

The whole farm SCCS calls for conversion of two or more adjacent fields into one contiguous area. An ideal example (Fig. 1) is illustrated with a two-year rotation of corn and winter wheat + stubble/cover crop. Both fields (one previously planted with corn, one previously planted with wheat) are converted into one large field. Both crops are now planted in the large field using a SCCS planting pattern. Management results could include the harvest of three crops: (1) winter wheat grain crop; (2) unprecedented tonnage stubble forage crop at grain harvest; (3) corn; and (4) continuous cover crop until planting the next corn crop.

In this example, the total rotational corn yield would be twice that of the conventional system. The stubble crop would produce an amazingly robust crop instead of one that was held back until small grain harvest allowed penetration of photosynthetically active incident sunlight, and the cover crop would extend for at least 18 months from corn harvest of year 1 until the next corn planting on that portion of the field. Wheat (or similar cool-season crop) will benefit from the additional exposure to incident sunlight to potentially produce a little more than expected if it were managed as a monoculture. Many other scenarios that include livestock integration could begin harvest as early as the corn reaches hard dough stage with the stubble forage crop as a good backdrop for, but not limited to, ruminants. Properly designed SCCS presented as guides for astute farmers to explore many options of their specific production systems will be described in more detail in later chapters.

The plan: a 360° community development protocol

An operational scenario for supporting the astute beginning farmer or small-acreage family farm to be successful in making a living offers tangible benefits of increased sustainable production, carbon sequestration, and net operating profits that not only sustain the family farm but also support the rural community. The sourcing production system can be illustrated using the corn–winter wheat, or soybean–winter wheat rotation, in a SCCS proposed by Deichman (2007) established on landscapes subject to significant soil erosion. Integration of ruminant livestock at stocking rates adequate for grazing corn stover and wheat during fall to early spring could be used to overcome local food deficits near so-called food deserts within the region. This system also offers entrepreneurial opportunities for underemployed rural residents including unemployed youth to aid in harvesting and preparing produce from selected floor crops included in the SCCS for local farmers markets and/or community-supported agriculture (CSA) enterprises. These options may be further supported through various community-based entities including consumer and producer cooperatives and employee-owned processors to aid in marketing farm-raised foods with a credit union option. Further details on potential impacts of the SCCS on community development are presented in Chapter 6, The economic and social conditions of the solar corridor cropping system (Hendrickson, 2019).

Why the SCCS was not discovered sooner?

Using the properly designed SCCS to test corn hybrids that typically produce high yields in a monoculture generally produce negative results, that is, yields would be lower or the hybrids would exhibit a negative response to the SCCS. Accordingly, most current popular commercial corn hybrids yield less in the SCCS, especially if only one hybrid is chosen to serve as a reliable predictor of corn's response to the SCCS. Thus, this offers a reasonable, if not incomplete, explanation of why the SCCS was not discovered sooner. However, splitting the SCCS treatments by hybrid provided hybrid-specific, positive results from a small portion of the pool of hybrids. That portion was about 10% based on seminal field trial results for north–south planted rows, somewhat less for east–west rows at sites near 40° North latitude (Deichman, 2000, 2009) Thus, alternative cropping systems such as the SCCS escaped discovery by mainstream research that focused on intensification of a single discipline featuring proprietary products, universal adaptations, or even a "preferred economic proprietary outcome." See Chapter 2, Variety selection for the solar corridor crop system (SCCS), for details on selection of the correct hybrids for the SCCS.

The proper development of the SCCS mandates that various solar corridor treatments are to be split by previously qualified hybrid and then by plant population, as deemed appropriate by agronomic expressions of the phenotype. The SCCS is designed to be more efficient in utilizing the key production input of incident sunlight, which defies economic valuation, that is, no price tag has been assigned to incident sunlight. Thus, the SCCS is a good example of being "disruptive" to modern corn production that evolved through modern society and technology, because, in part, it is contrary to current state-of-the-art production systems (Hoeft et al., 2000). The SCCS may (usually does) require more labor and specific equipment per land area unit, which may be rationally viewed as a multidimensional barrier to increasing farm size. Indeed, the SCCS does not require more land area to produce equivalent crop yields or net operating profits as typical of conventional production systems.

Summary

This chapter introduced the concept of the SCCS, an alternative system based on more efficient use of incident sunlight and atmospheric CO_2 for sustainable production of crops on less land than required for conventional production while improving soil and environmental quality. The development of the SCCS is in its early stages and, as will be noted in subsequent chapters, many important questions regarding research and practical applications remain to be answered. Additional advances in education, both in basic concepts of crop production and in extension, for providing optional production systems, in how benefits are realized by the agricultural and rural communities, are required if the SCCS is to be adopted in general practice. On a broader scale, deployment of SCCS may aid in meeting several global challenges (Clark, 2019) including reducing dependence on nonrenewable fossil fuels, reducing carbon generation of high levels of atmospheric CO_2, and providing locally accessible food for the underfed and undernourished local human population. Regardless, we believe that knowledge about efficient crop production and opportunities to adapt this knowledge in fields are adequately advanced to merit further development of the SCCS. It is our hope that this book provides the necessary information to proceed in planning strategies for deployment of the SCCS and in meeting global challenges to agricultural production.

Acknowledgments

Funding of all outlined seminal research was accomplished without any public or private company (customer paid product sourced) support. Deichman expresses his profound thanks to all (original and ongoing) collaborators and MRU, formed for the expressed purpose of completing the seminal research. Collaborative research with Kremer and Reinbott was partially funded by U.S.D.A., Agricultural Research Service.

C. LeRoy Deichman
Deichman Consulting, Shelbyville, MO, United States

Robert J. Kremer
University of Missouri, Columbia, MO, United States

References

Clark, K.M., 2019. The solar corridor crop system: a pivotal production system for African smallholder farmers. In: Deichman, C.L., Kremer, R.J. (Eds.), The Solar Corridor Crop System. Elsevier Academic Press, San Diego, CA (Chapter 7).

Crow, J.F., 1998. 90 years ago: the beginning of hybrid maize. Genetics 148, 923–928.

Deichman, C.L., 1980. Genotype by soil fertility interactions at .01%BLSD. In: Proceedings, National Fertilizer Solutions Assoc. 1980 Summer Round-Up, Chicago, IL.

Deichman, C.L., 2000. Plant Arrangement for Improving Crop Yields. U.S. Patent 6052,941 (date issued: 25.04.2000).

Deichman, C.L., 2002. Seminal corn yield results of the solar corridor corn production system III. Proceedings of International Annual Meetings, ASA-CSSA-SSSA, Indianapolis, IN.

Deichman, C.L., 2007. The solar corridor hypothesis. In: Proceedings of ASA–CSSA–SSSA Annual Meetings, New Orleans, LA, 4–8 November 2007. American Society of America, Madison, WI (paper no. 335-3).

Deichman, C.L., 2009. A sustainable farming system to maximize photosynthesis. In: Proceedings Farm System Design Symposium, Monterey, CA, 23–26 August 2009. Int. Environ. Model. Software Soc., Monterey, CA, pp. 87–88.

Deichman, C.L., 2019. Variety selection for the solar corridor crop system (SCCS). In: Deichman, C.L., Kremer, R.J. (Eds.), The Solar Corridor Crop System. Elsevier Academic Press, San Diego, CA (Chapter 2).

Deichman, C.L., Barrett, W., 1978. Heritability of corn hybrid response to plant nutrients. In: Proceedings of ASA–CSSA–SSSA Annual Meetings, Chicago, IL.

Deichman, C.L., Murrell, L., 1977. Differential corn hybrid responses to soil fertility variables. Proceedings of ASA-CSSA-SSSA Annual Meetings, Los Angeles, CA.

Deichman, C.L., and Warsaw, H., 1988. A novel research protocol to examine the source/sink relationship in corn. In: Proceedings of ASA–CSSA–SSSA Annual Meetings, Anaheim, CA.

Deichman, C.L., Watson, D., Wehunt, R.L., 1969. CFS Plan and Profit Corn Program (commercial publication).

Deichman, C.L., Nelson, R.A., Glanzer, D., 2001. Solar corridor corn yield results II. Agronomy Abstracts, ASA–CSSA–SSSA Annual Meetings, Charlotte, NC.

Deichman, C.L., Kremer, R.J., and Lupo, A., 2016. Over 1000 gallons of liquid biofuels per acre from four different corn hybrids on a Guilford fine sandy loam. In: Proceedings, International Annual Meetings, ASA–CSSA–SSSA, Phoenix, AZ.

Graham, E.R., Lopez, P.L., Dean, T.M., 1972. Artificial light as a factor influencing yields of high-population corn. Trans. ASABE 15, 576–579.

Hatfield, J., 2019. Photosynthesis and the solar corridor crop system. In: Deichman, C.L., Kremer, R.J. (Eds.), The Solar Corridor Crop System. Elsevier Academic Press, San Diego, CA (Chapter 1).
Hendrickson, M.K., 2019. The economic and social conditions of the solar corridor cropping system. In: Deichman, C.L., Kremer, R.J. (Eds.), The Solar Corridor Crop System. Elsevier Academic Press, San Diego, CA (Chapter 6).
Hoeft, R.G., Nafziger, E.D., Johnson, R.R., Aldrich, S.R., 2000. Modern Corn and Soybean Production. MCSP Publications, Champaign, IL.
Kremer, R.J., Deichman, C.L., 2014a. Introduction: the solar corridor concept. Agron. J. 106, 1817–1819.
Kremer, R.J., Deichman, C.L., 2014b. Soil quality and the solar corridor crop system. Agron. J. 106, 1853–1858.
Kremer, R.J., Deichman, C.L., Reinbott, T.M., Veum, K.S., 2017. The solar corridor cropping system for efficient gran production and improved soil quality on a claypan soil. Agron. Abstr. American Society of Agronomy, Madison, WI.
Liebman, M., Staver, C.P., 2001. Crop diversification for weed management. In: Liebman, M., Mohler, C.L., Staver, C.P. (Eds.), Ecological Management of Agricultural Weeds. Cambridge Univ. Press, Cambridge, UK, pp. 322–374.
Nelson, K.A., 2014. Corn yield response to the solar corridor in upstate Missouri. Agron. J. 106, 1847–1852.
Nielsen, R.L., 2017. Historical Corn Grain Yields for the U.S. <http://www.kingcorn.org/news/timeless/YieldTrends.html>.
Ottman, M.J., Welch, L.F., 1988. Supplemental radiation effects on senescence, plant nutriients, and yield of field-grown corn. Agron. J. 80, 619–626.
Ottman, M.J., Welch, L.F., 1989. Planting patterns and radiation interception, plant nutrient concentration, and yield in corn. Agron. J. 81, 167–174.
Pendleton, J.W., Egli, D.B., Peters, D.B., 1967. Response of *Zea mays* L. to a "light rich" field environment. Agron. J 59, 395–397.
Shull, G.H., 1908. The composition of a field of maize. Am. Breeders Assoc. Rep. 4, 296–301.
Smith, D., 2000. The solar corridor: patented 60" rows provide more sunlight, spike yields 40%. Farm Journal 124(9), special feature page 1.
Temesgen, A., Fakai, S., Rodriguez, D., 2017. Intercropping in sustainable maize cultivation. In: Watson, D. (Ed.), Achieving Sustainable Cultivation of Maize, Volume 2, Cultivation Techniques, Pest and Disease Control. Burleigh Dodds Science Publishing Ltd, Cambridge, UK, pp. 203–225.
Timlin, D.J., Fleisher, D.H., Kemanian, A.R., Reddy, V.R., 2014. Plant density and leaf area index effects on the distribution of light transmittance to the soil surface in maize. Agron. J. 106, 1828–1837.
Timlin, D.J., Kim, S.-H., Fleisher, D., Wang, Z., Reddy, V.R., 2019. Maize water use and yield in the solar corridor crop system: a simulation study. In: Deichman, C.L., Kremer, R.J. (Eds.), The Solar Corridor Crop System. Elsevier Academic Press, San Diego, CA (Chapter 3).

Further reading

Deichman, C.L., 2005. Floor crops for the solar corridor crop production system. In: Agronomy Abstracts, ASA–CSSA–SSSA Annual Meetings, Salt Lake City, UT.
Deichman, C.L., Nelson, R.A., 2000. Novel solar corridor corn yield results by four commercial hybrids I. In: Proceedings of ASA–CSSA–SSSA Annual Meetings, Minneapolis, MN.
Kremer, R.J., 2019. Soil health benefits of the solar corridor crop system. In: Deichman, C.L., Kremer, R.J. (Eds.), The Solar Corridor Crop System. Elsevier Academic Press, San Diego, CA (Chapter 4).

CHAPTER 1

Photosynthesis in the solar corridor system

Jerry L. Hatfield and Christian Dold
National Laboratory for Agriculture and the Environment, Ames, IA, United States

Contents

Introduction

Photosynthesis is the capture of light within the chloroplast that combines carbon dioxide (CO_2) with water (H_2O) to create a simple sugar ($C_6H_{12}O_6$) and can be expressed as

$$[6CO_2 + 6H_2O]\text{light} = C_6H_{12}O_6 + 6O_2 \tag{1.1}$$

However, this simple process is made complex by the factors that affect the availability of sunlight onto the leaves in a plant canopy. This can be envisioned through the light response curve diagram, which depicts how available sunlight (irradiance) is related to the rate of photosynthesis (Fig. 1.1). Important features in this diagram include the

The Solar Corridor Crop System
DOI: https://doi.org/10.1016/B978-0-12-814792-4.00001-2

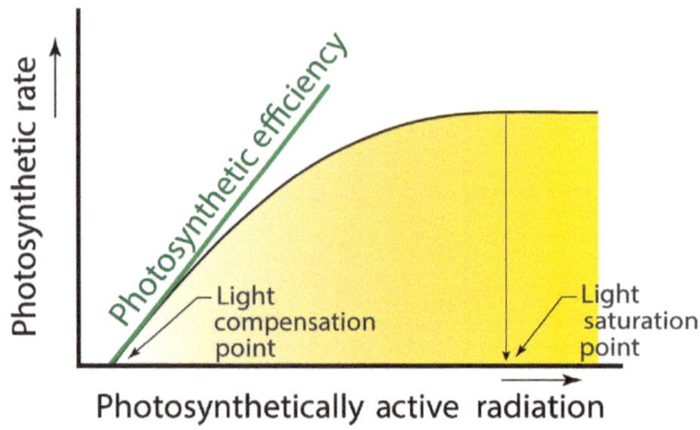

Figure 1.1 Generalized photosynthesis response to available photosynthetically active radiation showing light saturation level.

saturation point where there is no further change in photosynthetic rate with additional light, the compensation point where this amount of photosynthesis offsets the respiration rate, and the photosynthetic efficiency (or quantum yield) which is the rate of increase in photosynthesis per unit of light in the linear phase before saturation is reached. These principles of photosynthesis are known in basic plant physiology and determine the effectiveness of light capture within a plant canopy and the amount of photosynthesis generated (Marshall and Biscoe, 1980; Evans et al., 1993).

As described in Fig. 1.1, both the amount of incident radiation (1) and the efficiency of the plant to produce photosynthates (2) determine crop production as related to photosynthesis. Photosynthetic efficiency differs among plants with C_3, C_4, and crassulacean acid metabolism (CAM) pathway of carboxylation, and may be limited or enhanced by (un)favorable environmental factors, such as temperature (e.g., Ehleringer and Pearcy, 1983; Yamori et al., 2014), water (e.g., Chaves et al., 2002; Flexas et al., 2006; Ghannoum, 2009), ambient CO_2 concentration (e.g., Farquhar et al., 1980; Von Caemmerer and Furbank, 1999; Leakey et al., 2004; Ainsworth and Long, 2005; Sharkey et al., 2007), nutrients (e.g., Arnon, 1956; Huber, 1985; Sinclair and Horie, 1989; Abadía, 1992; Evans, 1996; Balakrishnan et al., 2000; Cakmak and Kirkby, 2008; Gu et al., 2017), and interactions among them, which eventually impact crop production.

Solar radiation is the sole source of energy for field crops to produce photosynthates, and light interception is often a limiting factor for crop production. For example, solar brightening (i.e., the decadal increase of incident solar radiation) has reportedly increased maize yields in the US Corn Belt, while solar dimming (i.e., the opposite effect) may have affected maize production in Northeast China (Tollenaar et al., 2017; Zhao et al., 2015). What becomes important in canopy level photosynthesis is the amount of light transmitted to each leaf within the plant canopy. This principle was first described by Monsi and Saeki (1953) where they applied the Beer–Bouguer–Lambert law to plant canopies as

$$I = I_o exp^{-k\mathrm{LAI}} \tag{1.2}$$

where I is the light intensity at any point in the canopy, I_o is the light at the top of the canopy, k is an extinction coefficient, and LAI is the leaf area index of the canopy above the level where I is estimated. Eq. (1.2) states that I decreases exponentially in plant canopies with increasing LAI and the angular arrangement of the leaves, expressed as the extinction coefficient (k). Note that Eq. (1.2) applies to random plant canopy structures with uniform radiation distribution at any canopy level. This is violated in field crops, where plant canopies are arranged in rows and leaves are clumped. In that case, a clumping index is added to the exponential term in Eq. (1.2). More complexity emerges when we consider that leaves are not all horizontal, light varies in both intensity and angle, and wind causes leaves to move. Values of k are primarily determined by the leaf angle, with canopies composed of large or horizontal leaves having higher k values than narrow and more vertical leaves showing lower values of k. Another form of k is expressed as

$$k = \frac{G(n_s)}{sin\beta} \tag{1.3}$$

where $G(n_s)$ is the projected area of the foliage and β is the solar elevation as derived by Warren Wilson (1965, 1967a). This form of the light penetration relationship introduces the concept that light penetration into a crop canopy will depend not only on the amount of foliage but also on the angle of the sun. For example, a dense canopy with a low sun angle will have light that is only on the upper leaves, while that same canopy at midday will have light penetrating to the lower leaves.

Several factors determine the distribution of radiation within a plant canopy, including distribution of direct and diffuse solar radiation, optical properties of the canopy, optical properties of the ground surface, and canopy architecture. One of the original experiments and analysis in the area of light distribution in plant canopies was conducted by Warren Wilson (1965, 1967a) using point quadrats to define the relationships among leaf area, leaf inclination, and sunlit leaf area at a range of sun angles.

These efforts led to the understanding that direct and diffuse light distributions within canopies are different, and penetration of direct solar radiation into plant canopies is governed by sun flecks. Ross (1975) thoroughly reviewed factors that determine how direct and diffuse solar radiation are distributed within plant canopies. Diffuse light and, more importantly, diffuse photosynthetically active radiation (PAR) are major components of light that penetrates into the plant canopy below the upper leaves, as light is scattered and reflected by leaves, plant surfaces, and soil. However, reported descriptions of diffuse light distribution within plant canopies are complex (Alados et al., 2002; Annandale et al., 2004; Wang et al., 2006).

Although diffuse radiation makes up the bulk part of transmitted light, the majority of incoming photons originate from direct radiation. Thereby, sun flecks with longer duration are more likely to reach full sun conditions than more frequent sun flecks with shorter duration. In addition, few big gaps contribute more to direct sunbeams than many small gaps (Chazdon et al., 1996; Körner, 2002).

The scientific literature has been extensive in the past decades on light distribution within canopies and models developed to describe the relationships between light and photosynthesis (Amthor, 1994, 2010; DeJong et al., 2011; dePury and Farquhar, 1997; Sellers et al., 1992; Stirling et al., 1994; Niinemets, 2010; Zhu et al., 2012; Alarcon and Sassenrath, 2015; Wang et al., 2017). These models provide a foundation for how light distribution within plant canopies is quantified and the potential for improving light capture by the complete canopy and not the upper leaves. Zhu et al. (2012) proposed that increasing the photosynthesis of the complete canopy will be key to increased productivity. This has been the premise underlying the solar corridor concept (Kremer and Deichman, 2014) as a variation of strip intercropping where two or more crops are placed in strips across the field with the potential for more efficient use of solar radiation, water, or nutrients. Deichman (2000) patented this concept applied to maize (*Zea mays* L.) production using either twin or single rows in very wide row widths. The overall concept is based on the premise of evaluating plant arrangements within a field to maximize the capture of solar radiation into the plant canopy. Throughout this chapter, we will focus on the factors that limit the efficient use of solar radiation through photosynthesis.

Radiation use efficiency

Radiation use efficiency (RUE) can be traced back to Wilson (1967b) as the relationship between the accumulation of biomass relative to the light intercepted by the crop. Loomis and Amthor (1999) stated that RUE would be a useful parameter to relate canopy photosynthesis to crop yield. The RUE is an important quantifier of crop production in relation to photosynthesis, as it combines both the amount of solar radiation captured by the crop and the efficiency of the crop to produce dry matter, assuming other factors are not limiting (Fig. 1.1). Measurements of RUE determined for a range of crops and different management systems involve the collection of dry biomass, grain yield or carbon assimilated, and the accumulation of intercepted PAR by the canopy over the life cycle of the crop. Long et al. (2006) suggested that canopy architecture would be one path toward increasing crop yield, which would focus on a more efficient conversion of available PAR into biomass or yield and directly related to factors contributing to improved RUE. Numerous experiments conducted around the world on RUE are summarized in Table 1.1, representing a wide range of environmental and agronomic practices. Loomis and Amthor (1999) proposed theoretical ranges for C_4 and C_3 plants showing the differences were large between these two physiological systems (Table 1.1). Their range of RUE for C_4 species was between 4.0 and $5.8\ g\ MJ^{-1}$ while for C_3 plants the range was $1.5-2.0\ g\ MJ^{-1}$. Values extracted from the literature show that actual measured values are often less than the theoretical values (Table 1.1).

The RUE approach has been utilized as a quantitative framework to determine how different agronomic systems intercept PAR and convert it into biomass or grain. For example, Bange et al. (1997) used shade cover to manipulate the incoming radiation environment and found RUE increased with an increase in diffuse radiation in sunflower (*Helianthus annuus* L.) canopy. This is similar to the results of Sinclair et al. (1992) who found that RUE increased with the amount of diffuse PAR in both maize and soybean. Another important aspect in evaluating RUE is the stage of growth of the plant and the form of the measurement. Dold et al. (2017b) showed a clear seasonal trend in RUE in maize, soybean, and natural prairie systems using eddy covariance derived datasets. Hatfield (2014) used a remote sensing method to obtain PAR interception values for four different crops and coupled these with measurements of total biomass or grain yield to estimate seasonal RUE. Rochette et al. (1995) found that RUE increased with LAI from 0.95 with LAI values less than 1 to 2.09 when LAI values were between 4 and 5. For the season, RUE was $1.80\ g\ MJ^{-1}$. However, if RUE was expressed in terms of CO_2 uptake the values ranged from $2.56\ g\ CO_2\ MJ^{-1}$ ($LAI < 1$) to

Table 1.1 Radiation use efficiency for different crops grown in different configurations.

Crop	Country	Year	System	Row width	Radiation use efficiency	References
C_4 species, theoretical					4.0–5.8 g MJ^{-1}	Loomis and Amthor (1999)
C_3 species, theoretical					1.5–2.0 g MJ^{-1}	Loomis and Amthor (1999)
Generic model focused on leek (*Allium porrion* L.) and celery (*Apium graveolens* L.)	Switzerland	1996, 1998	Ecophysiological model evaluation with intercropping of leak and celery	25–30 cm	Leek 2.28 ± 0.13 g MJ^{-1} Celery 2.17 ± 0.05 g MJ^{-1}	Baumann et al. (2002)
Alfalfa (*Medicago sativa* L.)	Argentina	2009–10	Spatial arrangement	10–30 cm	RUE 2.0 g MJ^{-1} for 15 cm spacing to 1.6 g MJ^{-1} for 30 cm spacing	Mattera et al. (2013)
Barley (*Hordeum vulgare* L.)	United States	2000–01	Spring barely comparison to wheat, seeding density	20 cm	0.90 ± 0.03–1.19 ± 0.03 g MJ^{-1} RUE decreased with increasing vapor pressure deficit	Kemanian et al. (2004)
Maize, grain	United States	2010–12	Comparison of tillage systems	76 cm	2.71–4.19 g MJ^{-1}	Hatfield (2014)
Maize	United States	1991–92	Maize with johnsongrass (*Sorghum halepense* L.)	69 cm	3.42–3.75 g MJ^{-1} maize alone 2.26–2.27 g MJ^{-1} johnsongrass alone 2.65 g MJ^{-1} mixed stand	Kiniry (1994)
Maize, grain	United States				3.5 ± 0.7 g MJ^{-1}	Kiniry et al. (1989)
Maize, grain	United States	2010–12	Seed orientation, plant population, and hybrids	76 cm	0.39–3.64 g MJ^{-1}	Torres et al. (2017)
Maize	United States	1998–2002	Plant population	76 cm	3.74–3.84 g MJ^{-1} no effect of population or years	Lindquist et al. (2005)

Maize, grain	Mexico	2008–10	Hybrids and plant population	80 cm	Plant population: 0.54–0.68 g MJ^{-1} Hybrids: 0.55–0.67 g MJ^{-1}	Morales-Ruiz et al. (2016)
Maize	China	2015–16	Cropping system and plant population	50–70 cm	0.90–1.69 g MJ^{-1}	Jia et al. (2018)
Maize, grain	China	2010–11	Hybrids	60 cm	1.6–2.7 g MJ^{-1}	Zhao et al. (2015)
Maize	Multiple sites				RUE increased with proportion of diffuse radiation from 1.5 (0.15 diffuse) to 2.0 g MJ^{-1} (> 0.4 diffuse)	Sinclair et al. (1992)
Cowpea, peas	Australia		Nonlimiting water		1.05 g MJ^{-1}	Muchow et al. (1993)
Grain sorghum (*Sorghum biocolor* (L.) Moench.)	United States		Two cultivars		1.7 and 2.2 g MJ^{-1} for the two cultivars	Hammer and Vanderlip (1989)
Grain sorghum	United States	1989	Cultivars and plant density at 5, 12, 19, and 26 plants m^{-2})	69 cm	3.46 ± 0.20 g MJ^{-1} with no difference among factors	Rosenthal et al. (1993)
Grain sorghum	United States	2009–10	Genotype comparison	76 cm	2.13 ± 0.33–3.53 ± 0.31 g MJ^{-1} in 2009 2.08 ± 0.35–3.83 ± 0.33 g MJ^{-1} in 2010, two genotypes with consistently higher RUE	Narayanan et al. (2013)
Mung bean, beans	Australia		Nonlimiting water		0.94 g MJ^{-1}	Muchow et al. (1993)
Rice, grain	United States				2.2 ± 0.2 g MJ^{-1}	Kiniry et al. (1989)
Rice	China	2012–13	Cultivar comparison	20 × 20 cm spacing	1.54–1.81 g MJ^{-1} with hybrid with higher leaf photosynthetic rate having higher RUE	Huang et al. (2016)

(*Continued*)

Table 1.1 (Continued)

Crop	Country	Year	System	Row width	Radiation use efficiency	References
Soybean, beans	Australia		Nonlimiting water		0.88 g MJ^{-1}	Muchow et al. (1993)
Soybean, beans	Multiple experiments				1.32–2.52 g MJ^{-1}	Sinclair and Muchow (1999)
Soybean, beans	United States	2010		76 cm	1.46 g MJ^{-1}	Singer et al. (2011)
Soybean, biomass	Canada	1992	Sample at different growth stages	18 cm	1.67–1.80 g MJ^{-1} for entire season	Rochette et al. (1995)
Soybean	United States	1997–99	Cultivars and plant density 4–90 plants m^{-2})	19 cm	Effect of plant density in decreasing RUE with an exponential relationship at R6 1.42 g MJ^{-1} at 12 plants m^{-2} to 1.05 at 134 plants m^{-2}	Purcell et al. (2002)
Soybean	Multiple sites				RUE increased with proportion of diffuse radiation from 1.0 (0.15 diffuse) to 1.4 g MJ^{-1} ($>$0.4 diffuse)	Sinclair et al. (1992)
Soybean, beans	United States	2010–12	Comparison of tillage systems	76 cm	1.15–1.76 g MJ^{-1} comparison of tillage systems	Hatfield (2014)
Sunflower, grain	United States				2.2 $\pm$ 0.3 g MJ^{-1}	Kiniry et al. (1989)
Sunflower, grain	Australia	1991–92	Radiation levels	75 cm	1.33–1.52 g MJ^{-1}	Bange et al. (1997)
Winter wheat	United States	2004–06	Seeding density with red clover interseeded	19 cm	No effect of plant density, ranged from 3.35 to 3.78 g MJ^{-1} Red clover RUE dependent on growth stage 1.40–2.52 g MJ^{-1}	Singer et al. (2007)

Triticale (*Triticum aestivum x Triticosecale* Wittmack)	United States	2004–06	Seeding density with red clover interseeded	19 cm	Seeding density effect, ranged from 2.84 to 3.40 g MJ^{-1} Red clover RUE dependent on growth stage 0.72–3.64 g MJ^{-1}	Singer et al. (2007)
Winter wheat	China	2008–10	Planting pattern and irrigation	Varied from 20 to 40 cm	RUE not affected by planting pattern, increased with irrigation	Han et al. (2013)
Wheat, field peaMustard	Australia, semiarid environment	1993–97	Rotation experiment	17.5 cm	Wheat 1.81 ± 0.05 g MJ^{-1} Field pea 1.52 ± 0.05 g MJ^{-1} Mustard 1.92 ± 0.12 g MJ^{-1} No detectable differences due to drought	O'Connell et al. (2004)
Forage rape, wheat, and forage sorghum	New Zealand	2010	N rates (0, 50, 100, 200 kg ha^{-1}) on C_3 dicot, C_3 monocot, and C_4 monocot	15 cm	Increased RUE with N rateWheat 1.30–1.66 g MJ^{-1} Forage rape 1.08–1.48 g MJ^{-1} Sorghum 1.60–2.12 g MJ^{-1}	Fletcher et al. (2013)
Winter wheat	China	2014–16	Planting configuration	15 cm	Uniform canopy had higher RUE at all plant densities, ranged from 1.45 ± 0.03 to 2.00 ± 0.07 g MJ^{-1}	Tao et al. (2018)
Wheat, grain	United States				2.8 ± 0.2 g MJ^{-1}	Kiniry et al. (1989)

3.31 g CO_2 MJ^{-1} ($4 < LAI < 5$) and the complete season RUE was 3.14 g CO_2 MJ^{-1} (Rochette et al., 1995).

Since plant biomass depends on nitrogen in nonlegume species and ample soil water, having water and nitrogen as nonlimiting factors increases RUE (Table 1.1). Plant population does not affect RUE in species that branch or tiller because the flexibility in growth habit promotes the efficient capture of light during the growing season. Results observed in different wheat experiments reveal that plant population or planting configuration is not a factor in RUE when other factors (water or nutrients) are not limiting growth (Table 1.1).

There are inconsistent differences in RUE with different cultivars or hybrids. For example, in grain sorghum (*Sorghum bicolor* [L.] Moench), no difference between two cultivars was observed by Rosenthal et al. (1993), but Hammer and Vanderlip (1989) found significant cultivar differences. Narayanan et al. (2013) found two of eight sorghum genotypes exhibited significantly different and higher RUE because of their ability to produce more biomass. Zhao et al. (2015) found significantly higher RUE in newer maize hybrids. Evaluation of RUE as a screening tool for genetic material and management systems is a viable technique because it is relatively easy to obtain accurate and reliable measurements of intercepted PAR over the growing season.

Photosynthesis in response to environmental factors

Light and [CO_2] are the primary factors that affect photosynthesis; however, several other factors also determine the efficiency of the photosynthetic process. Crops are subjected to a range of temperatures, varying water conditions, and nutrient availability during their growth cycle and each may limit the efficiency of the photosynthetic process. Long et al. (2006) suggested six factors that need to be addressed to overcome the limitations to photosynthesis. These included: (1) Rubisco with decreased oxygenase activity but without decreased catalytic rate; (2) efficient C_4 photosynthesis engineered into C_3 crops; (3) improved canopy architecture; (4) increased rate of recovery from photoprotection of photosynthesis; (5) introduction of higher catalytic rate foreign forms of Rubisco; and (6) increased capacity for regeneration of RuBP via overexpression of SbPas. Liu and Sun (2013) came to a similar conclusion after they evaluated limitations to photosynthesis in rice (*Orzya sativa* L.) and focused on the biochemical processes limiting photosynthesis. When we consider crop canopies, all factors that may potentially limit photosynthesis must be understood relative to the environment of the production system; however, of all potential factors, the manipulation of canopy architecture is most applicable to this discussion and could be achievable in a short period (Long et al., 2006).

Carbon dioxide response

Photosynthetic rates increase with the concentration of CO_2 in the atmosphere with a difference between C_3 plants [e.g., wheat (*Triticum aestivum* L.), soybean (*Glycine max* L. Merr.), and C_4 (e.g., maize)]. Plants with the C_3 metabolism show a continual increase in photosynthesis with CO_2 increase while C_4 plants do not exhibit a response to increasing $[CO_2]$s above a saturation point. The low response of C_4 plans was one reason why Leakey et al. (2004) concluded that the increase in atmospheric CO_2 will not contribute to increased yields in maize. Current atmospheric $[CO_2]$s are approximately 410 $\mu g\ g^{-1}$, which represent values at the top of a plant canopy. However $[CO_2]$ profiles within a plant canopy depend upon height and foliage density of the canopy and thereby affect $[CO_2]$-dependent rates of photosynthesis (Eq. 1.1). The exchange of wind within canopies will be an important factor affecting photosynthesis. If we assume that the efficiency of transport within canopies is similar to the extinction of light within canopies then we can expect that as photosynthesis extracts CO_2 from the air surrounding the leaves, replenishment of CO_2 into that volume will be a direct function of the exchange of momentum into the canopy. These profiles will change throughout the growing season depending upon foliage height and density. Profiles of CO_2 observed within maize at two growth stages, with maximum light interception, show that as the canopy continues to develop there is a more pronounced minimum in the $[CO_2]$ at midday because the air exchange becomes more limited and reduces the CO_2 exchange. During the night $[CO_2]$ increases within the canopy because of respiration and reduced momentum transfer causing the canopy concentrations within to exceed that in ambient air (Fig. 1.2). In maize canopies with an LAI of 2, there is incomplete canopy and extremely good mixing that maintained CO_2 profiles at a uniform concentration; however, as the canopy develops, mixing reduces and CO_2 depletes within the canopy during midday (Fig. 1.2). In soybean canopies, there is a different shape to these profiles during both night and midday because of the denser foliage (Fig. 1.3). With the

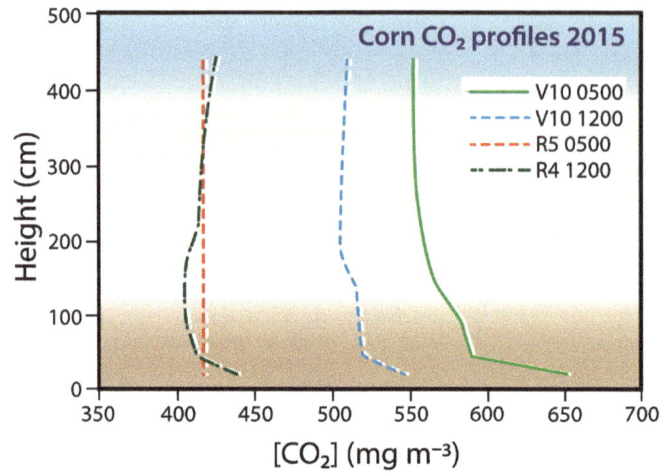

Figure 1.2 Profiles of CO_2 concentration in corn canopies at the V10 and R3 stages of development. Data collected in Central Iowa in 2015.

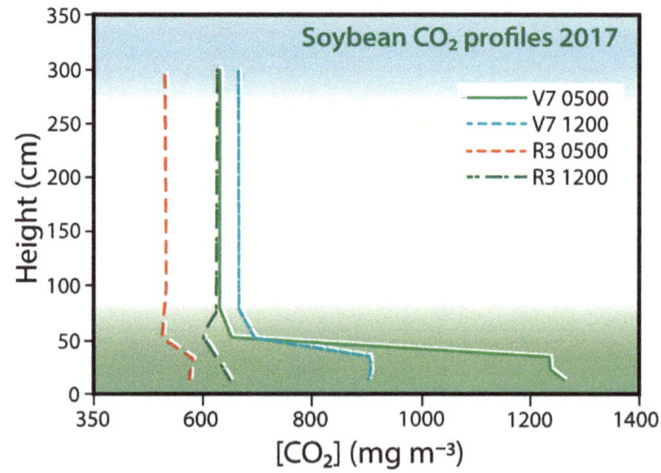

Figure 1.3 Profiles of CO_2 concentration in soybean canopies at the V8 and R6 stages of development. Data collected in Central Iowa in 2017.

greater foliage density, the minimum concentrations reveal that during the day, $[CO_2]$s reach a minimum but increase at night due to respiration from leaves. In both canopies, an interesting dynamic in the CO_2 concentrations during the day shows a minimum value at the height in the canopy where the light interception is maximum and the momentum transport is no longer sufficient to replenish the CO_2 supply. This effect is observed when crop canopies completely cover the soil and have maximum light interception. During the night, the concentration gradients show a large increase within the canopy due to respiration (Figs. 1.2 and 1.3). When crop canopies do not completely cover the soil, there is no evidence of a decrease in the CO_2 concentration.

Transport of CO_2 within plant canopies can be described in simple equation as

$$P = -\rho K_c \frac{dC}{dz} \quad (1.4)$$

where P is the flux density of CO_2 (g cm^{-2} s^{-1}), ρ is the density of air (g cm^{-3}), K_c is the diffusivity of CO_2 (cm^2 s^{-1}), and dC/dz is the vertical CO_2 gradient ($g_{CO_2}\, g_{air}^{-1}$ cm^{-1}). Gillespie (1971) and Gillespie and King (1971) utilized this approach to evaluate the gradients of CO_2, H_2O vapor, and sensible heat in maize canopies during the night. They found a minimum at the midheight of the canopy because of the lack of transport efficiency. An overview of the transfer of momentum and diffusivities in plant canopy was provided by Thom (1975), who detailed the processes that govern the exchange of energy and gases. Of critical importance to this discussion is that the dynamics of exchange of gases within a plant canopy are primarily determined by the stomatal resistance and the atmospheric resistance which we can describe in relation to CO_2 fluxes as

$$P = \rho \frac{[[CO_2]_x - [CO_2]_{x-1}]}{r_s + r_a} \quad (1.5)$$

where $[CO_2]_x$ is the concentration at some position x in the canopy and $[CO_2]_{x-1}$ is the concentration at position $x-1$, r_s is the stomatal resistance (s cm^{-1}) and r_a is the aerodynamic resistance (s cm^{-1}). Stomatal resistance determines the rate at which CO_2 enters the leaf and the aerodynamic resistance determines the rate of movement of CO_2 within the canopy volume. The latter factor becomes important when we begin to manipulate the arrangement of the canopy shape and the density of the leaves within the plant canopy. The arrangement of the canopy as changed by the concept of the solar corridor would have a positive major impact on the aerodynamic resistance of the canopy and would govern the rate of CO_2 exchange to the leaf. This will become a critical limiting factor as the $[CO_2]$ reaches a limit that reduces the photosynthetic rate of the leaves (Figs. 1.2 and 1.3, LAI).

Temperature

Temperature does not directly affect the photosynthetic process (Eq. 1.1); however, temperature changes are directly related to the enzymatic activity within the plant cell. Temperature response of enzymes exhibits an increase in activity until an optimum is reached when temperatures exceeding optimum markedly decrease enzyme activity. Note that plants can adapt to temperature optima according to changing ambient temperatures, thus reaching higher photosynthetic rates under new temperature regimes. C_3 plants are typically more adapted to cooler and temperate climates and have a wider range of optimum temperatures for photosynthesis, while C_4 and CAM plants are adapted to warm and (semi-)arid climates, respectively, both having lower optimum ranges (Yamori et al., 2014). Quantum yield of C_3 plants is reportedly temperature dependent, which is not the case for C_4 plants (Ehleringer and Pearcy, 1983). An important process at the canopy scale shows temperature changes throughout the growing season affects photosynthesis of plant stands during the year (Neiff et al., 2016). Stoy et al. (2014) found that the net ecosystem exchange (photosynthesis−respiration) was a function of the seasonal temperature patterns; however, this approach has not been applied to agriculture crops to determine the effect of temperature on canopy level photosynthesis. Bello et al. (1992) observed that nighttime temperatures affected RUE with higher temperatures reducing RUE because of increased respiration. Dold et al. (2017b) found a relationship between RUE and ambient air temperature for maize and natural prairie systems. A note of caution, however, is that the enzyme function within a leaf is determined by the actual temperature of the leaf and not air temperature.

It is a well-known fact that canopy temperatures, or more precisely foliage temperatures, are different from air. The upper leaf of the canopy where the light intensity is the greatest and PAR is not limiting to the photosynthetic process (Fig. 1.1) is the region of the canopy where the leaves will be cooler than air temperature. The evaporative cooling process decreases leaf temperature below air and would allow enzymatic reactions to be optimum for the plant. The differences in canopy and air temperature have been studied in various crops and indices developed to quantify the degree of crop stress. Dold et al. (2017a) applied the crop water stress index to canopies of maize, soybean, and native prairie and found that as the stress level and canopy temperatures increased, net ecosystem productivity decreased. Understanding that the temperatures of the foliage will change both temporally (diurnally and seasonally) and spatially (vertically within the canopy) will provide insights into the ability of the canopy to capture light and maintain an efficient photosynthetic process.

Water

Water deficits to the plant will directly affect the photosynthetic processes through two mechanisms: the direct effect on the physiological reactions within the cell and the effect upon stomatal resistance (Eq. 1.5). Leakey et al. (2006) showed that photosynthesis in maize was not affected by increasing [CO_2] with no water stress. However, stomatal resistance did not change with the elevated [CO_2] and no improved water use efficiency. Vanaja et al. (2017) in an experiment that combined increased air temperature by 3°C above ambient and imposed water deficits on maize decreased photosynthesis, increased stomatal resistance, decreased transpiration rate, and increased water use efficiency. In their study, effects of elevated temperature and water deficits were evident in the reproductive period of growth by reducing grain yield while plant biomass increased during the vegetative stage. The changes in photosynthesis were in proportion to the changes in stomatal resistance. Water deficits are major considerations in terms of limitations to the photosynthetic efficiency of crops (Tesfaye et al., 2006; Lu et al., 2015; Zhan et al., 2015). Ghannoum (2009) reviewed the literature and found water stress limited photosynthesis in C_3 and C_4 plants; however, the effect was more pronounced in C_4 plants despite the greater capacity in this photosynthetic pathway. Gonzalez-Dugo et al. (2010) point out the complex interactions between water stress and nitrogen management of crops. Understanding these interactions is a critical foundation for more effective crop management to sustain productivity required for future needs. The literature on water stress effects on plant growth and development is voluminous and beyond the scope of this chapter.

Nutrients

Availability of nutrients to plants and the effect on photosynthesis is related to leaf chlorophyll content. Gitelson et al. (2016) observed that vegetative productivity is related to absorption of light, which is directly proportional to leaf chlorophyll content. They introduced the concept of chlorophyll efficiency as this relationship:

$$\mathrm{ChlE} = \left(\frac{f\mathrm{APAR}_{\mathrm{green}}}{\mathrm{Chl}}\right) \times \mathrm{LUE} \tag{1.6}$$

where ChlE is the chlorophyll efficiency, *f*APAR is the absorbed PAR in the green canopy, Chl is the chlorophyll content, and LUE is the light use efficiency. The chlorophyll content could be obtained by any number of methods ranging from chlorophyll extraction from leaves to remote sensing methods. This concept of photosynthetic efficiency was proposed by Gebhardt (1986), described as

$$\eta_t + \eta_p + \eta_s + \eta_r + \eta_{ch} = \frac{\Delta G}{E} \tag{1.7}$$

where η_t is the thermodynamic efficiency, η_p is the trapping efficiency by the receptors sites, η_s is the spectral efficiency, η_r is the reflection of light by the leaves in the specific wavebands for photosynthesis, η_{ch} is the chemical efficiency, ΔG is the Gibbs free energy of the produced biomass, and E is the energy contained in the incident radiation. These factors provide insights into the complexities at the leaf level of the factors that determine how efficient the plant converts PAR into photosynthetic products. There are differences between C_4 and C_3 crops in the relationship between photosynthesis and N-use efficiency and determined by the lower cost for N in C_4 enzymes and the lower amount of Rubisco in the plant cells would lead to potentially higher N investment in thylakoid compounds. This would contribute to higher N-use efficiency for photosynthesis in C_4 plants (Makino et al., 2003). Tian et al. (2017) suggested that nitrogen fertilizer management in rice would enhance photosynthetic efficiency at the canopy level; however, these are not simple relationships and require an understanding of the interactions of crop physiology and agronomic practices.

Plant canopies affect light distribution and photosynthesis

Leaf area

Leaves as plant organs are the primary harvester of incident radiation. Leaf area and LAI (the ratio of leaf area to ground area) are therefore important parameters to quantify light distribution in canopies and photosynthesis (Eq. 1.2). The amount of PAR intercepted by the leaf area is a measure of the efficiency of the leaf area to capture incident radiation. Westgate et al. (2004) reported that most maize production systems intercept 59%–79% of PAR at maximum leaf area, while maximum grain yield is achieved at a 95% PAR interception. Leaf area and intercepted PAR in maize significantly changed with plant density, cropping system, hybrids, and nitrogen application (Timlin et al., 2014; Morales-Ruiz et al., 2016; Jia et al., 2018).

Another parameter to consider is the specific leaf area (SLA), which is the leaf area divided by the leaf dry matter, and is a measure of leaf thickness. Plants grown under low irradiance levels tend to have a higher SLA (i.e., thinner leaves) than under high irradiance levels, indicating that leaf structural components adapt to ambient radiation in order to capture more light. This adaptation strategy has reportedly high impact on maximizing carbon gain per unit leaf mass (Evans and Poorter, 2001).

Canopy architecture

Canopy architecture is the arrangement of leaves on a plant and the angular distribution of those leaves within the canopy on a horizontal and vertical plane, that is, leaf azimuth and inclination. The attenuation coefficient k (see Eq. 1.2) represents "the average projection of leaves on to a horizontal surface" (Campbell, 1986). Canopy architecture can be influenced by management practices, such as row spacing and plant population (plants m^{-2}) (e.g., Timlin et al., 2014). Seed orientation at planting reportedly impacts leaf azimuth (Torres et al., 2017; Fortin and Pierce, 1996). Row spacing and plant population determine the degree of mutual shading and amount of incident radiation entering the canopy; row position and plant population affect leaf azimuth (Fortin and Pierce, 1996) and attenuation coefficient (Maddonni et al., 2001) of maize. Part of the canopy architecture is related to how different leaves are shaded in the canopy and the distribution of diffuse radiation. Shade leaves are reportedly more planophile, while sun leaves are more erectophile, to either increase or decrease exposure to incident radiation (Niinemets, 2010). Li et al. (2010) found that shading in winter wheat improved growth and affected partitioning of dry matter into the grain. They attributed this response to improved light capture caused by a larger canopy leaf area, altered canopy architecture due to increased internode length, larger and thinner upper leaves, and increased pigment content, which led to increased RUE because of an increase in absorbed light due to more diffuse radiation and increased leaf chlorophyll b content (Li et al., 2010).

Leaf attenuation may also differ among plant hybrids, and hybrids may differ in their capability to adjust canopy architecture to incident radiation (Maddonni et al., 2001, 2002; Torres et al., 2017). Marchiori et al. (2010) observed that differences in three different sugarcane varieties (*Saccharum* spp.) were related to availability of light at the bottom of the canopy. The patterns and photosynthetic rates in the upper canopy were not different; however, the variety that had a different leaf inclination had greater photosynthesis in the lower leaves which improved vegetative growth due to increases in tillering, leaf area, and leaf and stalk biomass accumulation. Wang et al. (1995) observed a similar response in maize where canopies with more erect leaves (erectophile) compared to planophile leaves had a more uniform PAR and CO_2 distribution within the canopy. Maize with erect canopies exhibited increased grain yields (Wang et al., 1995).

Coops et al. (2017) found RUE was reduced at the top of the canopy rather than at the bottom, demonstrating the linkage of canopy architecture to RUE, similar to models developed by Birch et al. (2003). This suggests that canopy arrangement and plant architecture will be dominant factors affecting light capture and RUE. However, once the canopy develops to intercept the maximum amount of PAR, canopy architecture is no longer a factor (Maddonni et al., 2001).

Canopy architecture may impact the photosynthetic efficiency of plants. Evans et al. (1993) showed that vertical leaves had a lower light saturation point than horizontal leaves, indicating plant response to ambient light regimes. The canopy architecture not only affects the distribution of light and photosynthesis but also changes the microclimate that may affect the presence of insects or diseases. Pangga et al. (2013) reviewed the literature and found that the change in [CO_2] and the interacting effect between canopy architecture, crop growth within canopy microclimate, and potential insect and disease populations. In a similar analysis, Ney et al. (2013) found an interaction between the capture and use of natural resources, that is, water and light, and canopy architecture. Foliar diseases will affect leaf area, which in turn will reduce light interception by the canopy. Carretero et al. (2010) observed that distribution of disease incidence within a canopy will affect RUE in wheat. Changes in architecture will affect both the efficiency in use of light and water and directly determine whether the microclimate is potentially suitable for the support of insect or disease populations, and ultimately when the foliage is affected by insect or disease will determine the RUE.

Phenological stages

Photosynthesis depends upon the stage of development and is affected by the leaf age and leaf area. Canopy level photosynthesis will increase with increasing leaf area and greater light capture by the crop. Gitelson et al. (2014) found a strong relationship between canopy chlorophyll content and intercepted PAR to estimate gross primary productivity for maize and soybean. This concept was extended to evaluate the efficiency of the leaf chlorophyll and found that phenological stage of development was one of the primary factors affecting photosynthetic efficiency. The concept of "stay-green" traits in plants should lead to increased productivity and resistance to abiotic stress (Peleg et al., 2011; Gregersen et al., 2013; Reguera et al., 2013; Thomas and Ougham, 2014; Abdelrahman et al., 2017). This trait is vital to increased productivity; however, Thomas and Ougham (2014) found functional and nonfunctional "stay-green" traits. The functional traits maintain photosynthetic capacity and either delay the onset or slow the development of senescence. In plants with the nonfunctional traits, senescence proceeds with a decline in photosynthesis, but the leaf color is maintained because the degradation pathway of chlorophyll fails.

At the canopy level, photosynthesis will be governed by the ability of the canopy to intercept light. Interception will increase with leaf area (Eq. 1.2), and during the vegetative stage there will be an increase in leaf area coincident with phenological stage. In grain crops, the onset of grain filling will occur with the maximum leaf area, and during this period, photosynthesis will decline because of decreased green leaf area due to senescence. Leaf senescence occurs in the lower leaves first, and this allows for maintenance of the maximum amount of photosynthesis possible. The "stay-green" trait would be most beneficial during this stage because of the prolonged duration of effective leaf area. In a study on maize, Hatfield and Parkin (2014) related grain yield to the interception of PAR during the grain-filling period with management practices that promoted longer duration of increased PAR interception showing increased grain yields. Phenological stage is important to the efficiency of light capture in terms of leaf area and green leaf area duration.

Plant arrangement into canopies

The presentation and development of leaf area throughout the growing season and the canopy architecture will be determined by the arrangement of plants in a given area. This will be governed by the row spacing, the plant population (i.e., the distance between plants within a row), and plant composition.

Monocropping

In monocropping systems, altering plant population, row spacing, and row orientation can improve light distribution. Decreasing the row width in wheat during the season increases the leaf trajectory more rapidly leading to greater light interception and consequently high net carbon exchange rates (Wall and Kanemasu, 1990). In this study row widths were 18, 35, 53, and 71 cm and planted to maintain the same plant density per m^{-2}. Net carbon exchange rates decreased with increasing row width, and narrower row spacing canopies were able to maintain high net carbon exchange rates until leaf senescence decreased the leaf area below the threshold of maximum light interception ($LAI < 3.5$) (Wall and Kanemasu, 1990). A more recent study on maize comparing row configurations in China showed that narrow row canopies had a higher rate of photosynthesis and more uniform light distribution than the conventional planting (Liu and Song, 2012). They also observed that [CO_2] was lower in the narrow row canopies at the 50 cm strata similar to results shown in Fig. 1.3. Their planting configuration was 30 cm between rows separated by 170 cm and then two more rows, 36 cm between rows separated by 194 cm between rows, and an even distribution of 65 cm row spacing. The separation of the pairs of rows with the wider row spacing mimics the solar corridor concept and creates an environment conducive to diffuse PAR penetration into the lower canopy. The solar corridor concept is a maize monocropping or intercropping system where plants are grown in single or twin rows (twin row width: ~19 cm) with a between-row width of 152 cm and a recommended plant population of 39,500 plants ha^{-1} (Nelson, 2014; Deichman and Reinbott, 2018). This arrangement allows for better light distribution to leaves in the lower part of the canopy, similar to border row plants. However, grain yields were similar or lower than standard monocropping systems (Nelson, 2014). Part of this lack of yield response points out the need to consider hybrid selection for the solar corridor system (Deichman and Reinbott, 2018).

Models developed to quantify the potential interactions among plant shape, plant populations, and row width allow for quantitative assessment of leaf area distribution of plant canopies and distribution of

PAR within canopies. Stewart et al. (2003) described this approach for maize and found it was possible to accurately simulate PAR distribution within canopies and that an exponential relationship existed between LAI and canopy photosynthate production, canopy photosynthate ($g\ m^{-2}\ day^{-1}$) $= 47.42[1 - \exp(-0.8148LAI)]$ with a $R^2 = 0.73$, standard error of the estimate (SEE) $= 3.735\ g\ m^{-2}\ day^{-1}$. Any production factor that causes the LAI to increase more rapidly will increase the overall accumulation of photosynthate; they observed that maize canopies with more erect leaves exhibited greater canopy level photosynthate production because of the enhanced penetration of PAR into the canopy. Maddonni et al. (2001) observed that row spacing is only a factor when there is less than maximum leaf area to intercept the available PAR.

Changing row width has positive impacts on grain production. In mung bean [*Vigna radiata* (L.) Wilczek], Rachaputi et al. (2015) found that narrow rows (30 or 50 cm) compared to wide rows (90 or 100 cm) produced more biomass and greater yield. They extended this analysis to show that for 123 years in Australia, mung bean planted in narrow rows could produce up to 30% more grain in 95% of the growing seasons. The advantage of row width across all crops is in the more rapid accumulation of leaf area with the interception of more PAR being converted into photosynthate. We can hypothetically describe the relationship between LAI and photosynthetic rate as rapidly increasing with increasing LAI and reaching a plateau with LAI values above 4. There are differences between C_4 and C_3 plants in this response in the magnitude of the photosynthetic rates but not the shape of the relationship.

Intercropping

Mixtures of crops add complexity to the system; however, there have been assumptions that mixed stands could enhance the overall productivity of a unit area of land. Vance and Nevai (2007) proposed that the distribution of leaves in different canopy morphologies could be quantified by understanding the growth patterns of different species when placed in the same environment. They referred to this as a canopy partitioning model based on how the different plants grew and intercepted light. Intercropping represents a planned arrangement of crops on the same land area during the same time period. Several studies have shown the advantage of intercropping systems (Keating and Carberry, 1993; Ghaffarzadeh et al., 1994; Tournebize and Sinoquet, 1995; Lesoing and Francis, 1999;

Baumann et al., 2002; Jurik and Van, 2004; Seran and Brintha, 2010; Verdelli et al., 2012; Mao et al., 2014; Munz et al., 2014; El-Shamy et al., 2015; Husse et al., 2016; Gou et al., 2017) because of the altered light regime of the canopy. The major factor affected by mixtures of plants is the pattern of PAR interception into the canopies. The advantage to productivity in intercropping is due to the mixture of canopy heights to allow light to penetrate into the canopy at the border rows. In addition, intercropping takes advantage of the vertical distribution of crops to use the incident radiation resource effectively. Ghaffarzadeh et al. (1994) found a yield advantage in strip intercropping of maize and soybean in Iowa over a monoculture as long as the width of strip was minimal to prevent excessive shading of the soybean. In many cases, it is the shaded understory crop, which produces lower yields than in a monocropping setting. For example, while maize production was equal or higher in intercropping systems with smaller crops such as soybean (Verdelli et al., 2012), maize yields decreased in silver maple—maize intercropping systems due to tree shading (Miller and Pallardy, 2001). Yet, the combined yield of all crops is often higher than in monocropping, quantified as the land equivalent ratio (LER) (Gao et al., 2010). The LER can be calculated as (Mead and Willey, 1980)

$$\frac{\text{Yield}_{I\_C1}}{\text{Yield}_{M\_C1}} + \frac{\text{Yield}_{I\_C2}}{\text{Yield}_{M\_C2}} \tag{1.8}$$

where Yield_{I_C1} is the yield of intercropping system of first crop; Yield_{M_C1} is the yield of monocropping system of first crop; Yield_{I_C2} is the yield of intercropping system of second crop; and Yield_{M_C2} is the yield of monocropping system of second crop.

Intercropping systems with LER values >1 are considered advantageous to monocropping systems.

The solar corridor system can be used as an intercropping system, tested by Nelson (2014) using corn-wheat intercrops. Both wheat and corn were lower than in monocropping settings. Using Eq. (1.8), LER <1 was calculated with crop proportions of 0.27 for corn and 0.57 for wheat. These studies did not consider that the exchange of CO_2 caused by wind flow within canopies of varying shape and size could be beneficial in maintaining the [CO_2] at nonlimiting levels within the canopy volume (Figs. 1.2 and 1.3).

Increasing plant productivity

Plant level

Increasing photosynthetic rates at the plant level should focus on two primary factors: (1) exposing leaves to PAR levels that are above light saturation values (Fig. 1.1) and (2) maintaining environmental conditions so that light capture for the photosynthetic process functions at the optimum level. Syntheses of these factors detailed by Long et al. (2006) and Zhu et al. (2012) outline the potential changes that could be made at the leaf and canopy level. We can assume each plant is operating at maximum efficiency, and to achieve increases in productivity we have to focus on how we alter the canopy for maximum photosynthesis.

Melis (2009) proposed, at the leaf level, increasing the efficiency of conversion of solar energy into photosynthesis would yield positive results. Similarly Tholen et al. (2012) proposed the ideal leaf anatomical features for different environments. In warm temperate environments with high light and adequate soil water, they proposed:

> *Thick leaves with many thin chloroplasts allow more light absorption in a dense canopy and enlarge the surface area available for CO_2 diffusion. A high stomatal density on both sides of the leaf facilitates CO_2 diffusion. Thin cell walls and small mesophyll cells further reduce diffusion resistances. Elongated palisade cells are necessary for a homogeneous light distribution within a leaf. Erect leaves allow for more leaves in a dense canopy, no specific palisade cells are necessary and all cells can be thin walled, heavily lobed and small resulting in a large surface area for diffusion. The erect leaves are usually thin and have a low heat capacity. Without high transpiration rates such leaves can become too hot under direct sunlight, therefore a high hydraulic conductance is necessary for maintaining a high stomatal conductance. This further requires a high vein density, large veins and bundle-sheath extensions. Extended lignification of the vasculature and bundle-sheath extension are necessary to maintain leaf shape.*

In contrast, in warm, high-light, semiarid environments:

> *the ideal leaf structure constrained by water relations. More stomata on the lower side of the leaf. The leaf may be somewhat thinner than under wet conditions to minimize CO_2 diffusion, but this might lead to more water losses. Thick cell walls are necessary to maintain the cell structure under dry conditions. Elongated palisade cells homogenize the light within the leaf. A C_4 metabolism and anatomy would be most beneficial under these conditions. (Tholen et al., 2012)*

While in cool environments with lower or more diffuse lights Tholen et al. (2012) proposed, "Thin leaves are sufficient to capture all available light. Occasional periods of frost require small veins. Cells should be small and possibly lobed to increase the surface area available for diffusion."

Canopy level

The arrangement of plants into a canopy determines the effectiveness of a group of plants in intercepting and distributing PAR throughout the canopy. The relationship between LAI and photosynthetic rate governs the impact of row width, plant population, plant architecture, and crop stress due to water or nutrients. De Bossoreille de Ribou et al. (2013) outlined multiple factors to be considered for increasing productivity. Following the analysis of Zhu et al. (2008) the maximum theoretical conversion efficiency of solar energy to biomass is 4.6% for C_3 photosynthesis at 30°C and 380 $\mu g\, g^{-1}$ atmospheric CO_2, and 6% for C_4 photosynthesis. De Bossoreille de Ribou et al. (2013) proposed the following factors as valuable for improving productivity: plant architecture, indeterminate growth, vegetative/reproductive transition, sink/source flow relationships, extensive secondary growth in perennials, water and nutrient use efficiency, light interception efficiency, C_3/C_4 photosynthetic pathways, and stress avoidance processes. However, not all hybrids may be able to respond to improved light regimes and plant arrangements. Deichman and Reinbott (2018) state that most of the newer maize hybrids lacked a "flex-ear" response, that is, hybrids with increased ear growth according to the spatial plant arrangement. This means that a fundamental change in cropping systems, such as the solar corridor concept, would require plant hybrids with optimal response to these changes. We have the potential to increase productivity through manipulation of crop canopies to more efficiently capture light and maintain CO_2 levels throughout the canopy to be nonlimiting. We need to address these remaining challenges to ensure we can continue to enhance the productivity of all crops.

Conclusion

Plant productivity is dependent upon the photosynthetic process and the efficiency of conversion of sunlight into various products including yield. There are two approaches to use to increase the efficiency of this process at both the leaf levels, and it has been suggested that there are several alterations that could potentially impact productivity. The largest benefit for productivity is at the canopy level where we can begin to arrange leaves into canopies. The overall photosynthetic rate is determined by the accumulation of leaf area and the arrangement of the leaves into canopies for the interception of light and the distribution of light through the canopy. There are multiple components of canopy level photosynthesis, light interception, CO_2 distribution and gradients within canopies, and plant water status. The results from previous studies on canopy arrangements including row spacing in monocultures, intercropping with multiple crops, and changing plant populations with row spacing have shown the potential for enhancing productivity. Central to this effort is the focus on RUE for different cropping systems and then coupling this with water and nutrient use efficiency.

The increase in RUE by different cropping systems offers the potential to increase productivity for a given unit of water and light. Understanding the principles described in this chapter and utilizing these to help design new and improved cropping systems such as the solar corridor will increase our overall productivity and achieve the goal of increased food security to meet the population needs for food in the remainder of this century.

References

Abadía, J., 1992. Leaf responses to Fe deficiency: a review. J. Plant Nutr. 15 (10), 1699−1713.

Abdelrahman, M., El-Sayed, M., Jogaiah, S., Burritt, D.J., Tran, L.P., 2017. The "STAY-GREEN" trait and phytohormone signaling networks in plants under heat stress. Plant Cell Rep. 36, 1009−1025. Available from: https://doi.org/10.1007/s00299-017-2119-y.

Ainsworth, E.A., Long, S.P., 2005. What have we learned from 15 years of free-air CO_2 enrichment (FACE)? A meta-analytic review of the responses of photosynthesis, canopy properties and plant production to rising CO_2. New Phytol. 165 (2), 351−372.

Alados, I., Foyo-Moreno, I., Olmo, F.J., Alados-Arboledas, L., 2002. Improved estimation of diffuse photosynthetically active radiation using two spectral models. Agric. For. Meteorol. 111, 1−12.

Alarcon, V.J., Sassenrath, G.F., 2015. Optimizing canopy photosynthetic rate through PAR modeling in cotton (*Gossypium* spp.) crops. Comput. Electron. Agric. 119, 142−152. Available from: https://doi.org/10.1016/j.compag.2015.10.010.

Amthor, J.S., 1994. Scaling CO_2 photosynthesis relationships from the leaf to the canopy. Photosynth. Res. 39, 321−350.

Amthor, J.S., 2010. From sunlight to phytomass: on the potential efficiency of converting solar radiation to phyto-energy. New Phytol. 188, 939−959.

Annandale, J.G., Jovanovic, N.Z., Campbell, G.S., Sautoy, N.D., Lobit, P., 2004. Two-dimensional solar radiation interception model for hedgerow fruit trees. Agric. For. Meteorol. 121, 207−225.

Arnon, D.I., 1956. Phosphorus metabolism and photosynthesis. Annu. Rev. Plant Physiol. 7 (1), 325−356.

Balakrishnan, K., Rajendran, C., Kulandaivelu, G., 2000. Differential responses of iron, magnesium, and zinc deficiency on pigment composition, nutrient content, and photosynthetic activity in tropical fruit crops. Photosynthetica 38 (3), 477−479.

Bange, M.P., Hammer, G.L., Rickert, K.G., 1997. Effect of radiation environment on radiation use efficiency and growth of sunflower. Crop Sci. 37, 1208–1214.

Baumann, D.T., Bastiaans, L., Goudriaan, J., van Laar, H.H., Kropff, M.J., 2002. Analysing crop yield and plant quality in an intercropping system using an eco-physiological model for interplant competition. Agric. Syst. 73, 173–203.

Bello, M.J., Wright, G.C., Hammer, G.L., 1992. Night temperature affects radiation-use efficiency in peanut. Crop Sci. 32, 1329–1335.

Birch, C.J., Andrieu, B., Fournier, C., Vos, J., Room, P., 2003. Modeling kinetics of plant canopy architecture: concepts and applications. Eur. J. Agron. 19, 519–533. Available from: https://doi.org/10.1016/S1161-0301(02)00183-1.

Cakmak, I., Kirkby, E.A., 2008. Role of magnesium in carbon partitioning and alleviating photooxidative damage. Physiol. Plant. 133 (4), 692–704.

Campbell, G.S., 1986. Extinction coefficients for radiation in plant canopies calculated using an ellipsoidal inclination angle distribution. Agric. For. Meteorol. 36 (4), 317–321.

Carretero, R., Serrago, R.A., Banca, M.O., Perelló, A.E., Miralles, D.J., 2010. Absorbed radiation and radiation use efficiency as affected by foliar diseases in relation to their vertical position into the canopy in wheat. Field Crops Res. 116, 184–195. Available from: https://doi.org/10.1016/j.fcr.2009.12.009.

Chaves, M.M., Pereira, J.S., Maroco, J., Rodrigues, M.L., Ricardo, C.P.P., OsÓRio, M. L., et al., 2002. How plants cope with water stress in the field? Photosynthesis and growth. Ann. Bot. 89 (7), 907–916.

Chazdon, R.L., Pearcy, R.W., Lee, D.W., Fetcher, N., 1996. Photosynthetic responses of tropical forest plants to contrasting light environments. In: Chazdon, R.L., Mulkey, S. S., Smith, A.P. (Eds.), Tropical Forest Plant Ecophysiology. Chapman & Hall, London, UK (Chapter 1; ISBN: 0-412-03571-5675).

Coops, N.C., Hermosilla, T., Hilker, T., Black, T.A., 2017. Linking stand architecture with canopy reflectance to estimate vertical patterns of light-use efficiency. Remote Sens. Environ. 194, 322–330. Available from: https://doi.org/10.1016/j.rse.2017.03.025.

Deichman, C. LeRoy, 2000. Plant arrangement for improving crop yields. US Patent 6052,941. Patent issued 23 April 2000.

Deichman, C. LeRoy, Reinbott, T., 2018. Know your community: the solar corridor crop system. Crops Soils Agron. News 63 (3), 19–21.

DeJong, T.M., Da Silva, D., Vos, J., Escobar-Gutierrez, A.J., 2011. Using functional-structural plant models to study, understand and integrate plant development and eco-physiology. Ann. Bot. 108, 987–989.

dePury, D.G.G., Farquhar, G.D., 1997. Simple scaling of photosynthesis from leaves to canopies without the errors of big-leaf models. Plant Cell Environ. 20, 537–557.

de Bossoreille de Ribou, S., Douam, F., Hamant, O., Frohlich, M.W., Negrutiu, I., 2013. Plant science and agricultural productivity: why are we hitting the yield ceiling? Plant Sci. 210, 159–176. Available from: https://doi.org/10.1016/j.plantsci.2013.05.010.

Dold, C., Büyükcangaz, H., Rondinelli, W., Prueger, J., Sauer, T., Hatfield, J., 2017b. Long-term carbon uptake of agro-ecosystems in the Midwest. Agric. For. Meteorol. 232, 128–140.

Dold, C., Hatfield, J.L., Prueger, J., Sauer, T., Büyükcangaz, H., Rondinelli, W., 2017a. Long-term application of the crop water stress index in midwest agro-ecosystems. Agron. J. 109, 2172–2181. Available from: https://doi.org/10.2134/agronj2016.09.0494.

Ehleringer, J., Pearcy, R.W., 1983. Variation in quantum yield for CO_2 uptake among C_3 and C_4 plants. Plant Physiol. 73 (3), 555–559.

El-Shamy, M.A., Abdel-Wahab, T.I., Abdel-Wahab, S.I., Ragheb, S.B., 2015. Advantages of intercropping soybean with maize under two maize plant distributions and three mineral nitrogen fertilizer rates. Adv. Biosci. Bioeng. 3, 30–48. Available from: https://doi.org/10.11648/j.abb.20150304.11.

Evans, J., Poorter, H., 2001. Photosynthetic acclimation of plants to growth irradiance: the relative importance of specific leaf area and nitrogen partitioning in maximizing carbon gain. Plant Cell Environ. 24 (8), 755–767.

Evans, J.R., 1996. Developmental constraints on photosynthesis: effects of light and nutrition. In: Baker, N.R. (Ed.), Photosynthesis and the Environment. Springer, Berlin, Germany, pp. 281–304.

Evans, J.R., Jakobsen, I., Ögren, E., 1993. Photosynthetic light-response curves. Planta 189 (2), 191–200.

Farquhar, G., von Caemmerer, S.V., Berry, J., 1980. A biochemical model of photosynthetic CO_2 assimilation in leaves of C_3 species. Planta 149 (1), 78–90.

Fletcher, A.L., Johnstone, P.R., Chakwizira, E., Brown, H.E., 2013. Radiation capture and radiation use efficiency in response to N supply for crop species with contrasting canopies. Field Crops Res. 150, 126–134. Available from: https://doi.org/10.1016/j.fcr.2013.06.014.

Flexas, J., Bota, J., Galmés, J., Medrano, H., Ribas-Carbó, M., 2006. Keeping a positive carbon balance under adverse conditions: responses of photosynthesis and respiration to water stress. Physiol. Plant. 127 (3), 343–352.

Fortin, M.C., Pierce, F.J., 1996. Leaf azimuth in strip-intercropped corn. Agron. J. 88 (1), 6–9.

Gao, Y., Duan, A., Qiu, X., Sun, J., Zhang, J., Liu, H., et al., 2010. Distribution and use efficiency of photosynthetically active radiation in strip intercropping of maize and soybean. Agron. J. 102 (4), 1149–1157.

Gebhardt, W., 1986. Photosynthetic efficiency. Radiat. Environ. Biophys. 25, 275–288.

Ghaffarzadeh, M., Prechac, F.G., Cruse, R.M., 1994. Grain yield response of corn, soybean, and oat grown in a strip intercropping system. Am. J. Alternative Agric. 9, 171–177.

Ghannoum, O., 2009. C_4 photosynthesis and water stress. Ann. Bot. 103, 635–644. Available from: https://doi.org/10.1093/aob/mcn093.

Gillespie, T.J., 1971. Carbon dioxide profiles and apparent diffusivities in corn fields at night. Agric. Meteorol. 8, 51–57.

Gillespie, T.J., King, K.M., 1971. Night-time sink strengths and apparent diffusivities within a corn canopy. Agric. Meteorol. 8, 59–67.

Gitelson, A.A., Peng, Y., Arkebauer, T.J., Schepers, J., 2014. Relationships between gross primary production, green LAI, and canopy chlorophyll content in maize: implications for remote sensing of primary production. Remote Sens. Environ. 144, 65–72. Available from: https://doi.org/10.1016/j.rse.2014.01.004.

Gitelson, A.A., Peng, Y., Viña, A., Arkebauer, T., Schepers, J.S., 2016. Efficiency of chlorophyll in gross primary productivity: a proof of concept and application in crops. J. Plant Physiol. 210, 101–110. Available from: https://doi.org/10.1016/j.jplph.2016.05.019.

Gonzalez-Dugo, V., Durand, J.-L., Gastal, F., 2010. Water deficit and nitrogen nutrition of crops. A review. Agron. Sustain. Dev. 30, 529–544. Available from: https://doi.org/10.1051/agro/2009059.

Gou, F., van Ittersum, M.K., Simon, E., Leffelaar, P.A., van der Putten, P.E.L., Zhangc, L., et al., 2017. Intercropping wheat and maize increases total radiation interception and wheat RUE but lowers maize RUE. Eur. J. Agron. 84, 125–139. Available from: https://doi.org/10.1016/j.eja.2016.10.014.

Gregersen, P.L., Culetic, A., Boschian, L., Krupinska, K., 2013. Plant senescence and crop productivity. Plant Mol. Biol. 82, 603–622.

Gu, J., Chen, Y., Zhang, H., Li, Z., Zhou, Q., Yu, C., et al., 2017. Canopy light and nitrogen distributions are related to grain yield and nitrogen use efficiency in rice. Field Crop Res. 206, 74–85. Available from: https://doi.org/10.1016/j.fcr.2017.02.021.

Hammer, G.L., Vanderlip, R.L., 1989. Genotype-by-environment interaction in grain sorghum: I. Effects of temperature on radiation use efficiency. Crop Sci. 29, 370–376.

Han, Y.Y., Wang, G.Y., Zhou, X.B., Chen, Y.H., Liu, P., 2013. Radiation use efficiency and yield response of winter wheat to planting patterns and irrigation in Northern China. Agron. J. 106, 168–174. Available from: https://doi.org/10.2134/agronj2013.0252.

Hatfield, J.L., 2014. Radiation use efficiency: evaluation of cropping and management systems. Agron. J. 106, 1820–1827. Available from: https://doi.org/10.2134/agronj2013.0310.

Hatfield, J.L., Parkin, T.B., 2014. Enhanced efficiency fertilizers: effect on agronomic performance of corn in Iowa. Agron. J. 106, 771–780. Available from: https://doi.org/10.2134/agronj2013.0104.

Huang, M., Shan, S., Zhou, X., Chen, J., Cao, F., Jang, L., et al., 2016. Leaf photosynthetic performance related to higher radiation use efficiency and grain yield in hybrid rice. Field Crops Res. 193, 87–93. Available from: https://doi.org/10.1016/j.fcr.2016.03.009.

Huber, S.C., 1985. Role of potassium in photosynthesis and respiration. In: Munson, R. D. (Ed.), Potassium in Agriculture. ASA, CSSA, SSSA, Madison, WI, pp. 369–396. Available from: http://doi.org/10.2134/1985.

Husse, S., Huguenin-Elie, O., Buchmann, N., Lüscher, A., 2016. Larger yields of mixtures than monocultures of cultivated grassland species match with asynchrony in shoot growth among species but not with increased light interception. Field Crops Res. 194, 1–11. Available from: https://doi.org/10.1016/j.fcr.2016.04.021.

Jia, Q., Sun, L., Mou, H., Ali, S., Liu, D., Zhang, Y., et al., 2018. Effects of planting patterns and sowing densities on grain-filling, radiation use efficiency and yield of maize (*Zea mays* L.) in semi-arid regions. Agric. Water Manag. 201, 287–298.

Jurik, T.W., Van, K., 2004. Microenvironment of a corn–soybean–oat strip intercrop system. Field Crops Res. 90, 335–349.

Keating, B.A., Carberry, P.S., 1993. Resource capture and use in intercropping: solar radiation. Field Crops Res. 34, 273–301.

Kemanian, A.R., Stöckle, C.O., Huggins, D.R., 2004. Variability of barley radiation-use efficiency. Crop Sci. 44, 1662–1672.

Kiniry, J.R., 1994. Radiation-use efficiency and grain yield of maize competing with Johnsongrass. Agron. J. 86, 554–557.

Kiniry, J.R., Jones, C.A., O'Toole, J.C., Blanchet, R., Cabelguenne, M., Spanel, D.A., 1989. Radiation-use efficiency in biomass accumulation prior to grain-filling for five grain-crop species. Field Crops Res. 20, 51–64. Available from: https://doi.org/10.1016/0378-4290(89)90023-3.

Körner, C., 2002. Pflanzen im Lebensraum. Straßburger, Lehrbuch der Botanik für Hochschulen. Spektrum Akademischer Verlag, Heidelberg, Berlin, p. 1123 (Chapter 13, ISBN: 3-8274-1010-X).

Kremer, R.J., Deichman, C. LeRoy, 2014. Introduction: the solar corridor concept. Agron. J. 106, 1817–1819. Available from: https://doi.org/10.2134/agronj14.0291.

Leakey, A., Bernacchi, C., Dohleman, F., Ort, D., Long, S., 2004. Will photosynthesis of maize (*Zea mays*) in the US Corn Belt increase in future [CO_2] rich atmospheres? An analysis of diurnal courses of CO_2 uptake under free-air concentration enrichment (FACE). Global Change Biol. 10 (6), 951–962.

Leakey, A.D.B., Uribelarrea, M., Ainsworth, E.A., Naidu, S.L., Rogers, A., Ort, D.R., et al., 2006. Photosynthesis, productivity, and yield of maize are not affected by open-air elevation of CO_2 concentration in the absence of drought. Plant Physiol. 140, 779–790. Available from: https://doi.org/10.1104/pp.105.073957.

Lesoing, G.W., Francis, C.A., 1999. Strip intercropping effects on yield and yield components of corn, grain sorghum, and soybean. Agron. J. 91, 807–813.

Li, H., Jiang, D., Wollenweber, B., Dai, T., Cao, W., 2010. Effect of shading on morphology, physiology, and grain yield of winter wheat. Europ. J. Agron. 33, 267–275. Available from: https://doi.org/10.1016/j.eja.2010.07.002.

Lindquist, J.L., Arkebauer, T.J., Walters, D.T., Cassman, K.G., Dobermann, A., 2005. Maize radiation use efficiency under optimal growth conditions. Agron. J. 97, 72–78.

Liu, T.D., Song, F.B., 2012. Maize photosynthesis and microclimate within the canopies at grain-filling stage in response to narrow-wide row planting patterns. Photosynthetica 50, 215–222. Available from: https://doi.org/10.1007/s11099-012-0011-0.

Liu, Z., Sun, N., 2013. Enhancing CO_2 photosynthetic use efficiency in rice: approaches and challenges. Acta Physiol. Plant. 35, 1001–1009. Available from: https://doi.org/10.1007/s11738-012-1171-z.

Long, S.P., Zhu, X., Naidu, S.L., Ort, D.R., 2006. Can improvement in photosynthesis increase crop yields? Plant Cell Environ. 29, 315–330. Available from: https://doi.org/10.1111/j.1365-3040.2005.01493.x.

Loomis, R.S., Amthor, J.S., 1999. Yield potential, plant assimilatory capacity, and metabolic efficiencies. Crop Sci. 39, 1584–1596. Available from: https://doi.org/10.2135/cropsci1999.3961584x.

Lu, H.B., Qiao, Y.M., Gong, X.C., Li, H.Q., Zhang, Q., Zhao, Z.H., et al., 2015. Influence of drought stress on the photosynthetic characteristics and dry matter accumulation of hybrid millet. Photosynthetica 53, 306–311. Available from: https://doi.org/10.1007/s11099-015-0120-7.

Maddonni, G.A., Otegui, M.E., Cirilo, A.G., 2001. Plant population density, row spacing, and hybrids effects on maize canopy architecture and light attenuation. Field Crops Res. 71, 183–193.

Maddonni, G.A., Otegui, M.E., Andrieu, B., Chelle, M., Casal, J.J., 2002. Maize leaves turn away from neighbors. Plant Physiol. 130 (3), 1181–1189.

Makino, A., Sakuma, H., Sudo, E., Mae, T., 2003. Differences between maize and rice in N-use efficiency for photosynthesis and protein allocation. Plant Cell Physiol. 44, 952–956.

Mao, L., Zhang, L., Zhao, X., Liu, S., van der Werf, W., Zhang, S., et al., 2014. Crop growth, light utilization and yield of relay intercropped cotton as affected by plant density and a plant growth regulator. Field Crops Res. 155, 67–76. Available from: https://doi.org/10.1016/j.fcr.2013.09.021.

Marshall, B., Biscoe, P.V., 1980. A model for C_3 leaves describing the dependence of net photosynthesis on irradiance. J. Exp. Bot. 31 (1), 29–39.

Marchiori, P.E.R., Ribeiro, R.V., DaSilva, L., Machado, R.S., Machado, E.C., Scarpari, M.S., 2010. Plant growth, canopy photosynthesis and light availability in three sugarcane varieties. Sugar Tech. 12, 160–166.

Mattera, J., Romero, L.A., Cuatrín, A.L., Cornaglia, P.S., Grimold, A.A., 2013. Yield components, light interception and radiation use efficiency of Lucerne (*Medicago sativa* L.) in response to row spacing. Eur. J. Agron. 45, 87–95. Available from: https://doi.org/10.1016/j.eja.2012.10.008.

Mead, R., Willey, R.W., 1980. The concept of a "land equivalent ratio" and advantages in yields from intercropping. Exp. Agric. 16 (3), 217–228.

Melis, A., 2009. Solar energy conversion efficiencies in photosynthesis: minimizing the chlorophyll antennae to maximize efficiency. Plant Sci. 177, 272–280. Available from: https://doi.org/10.1016/j.plantsci.2009.06.005.

Miller, A.W., Pallardy, S.G., 2001. Resource competition across the crop-tree interface in a maize-silver maple temperate alley cropping stand in Missouri. Agrofor. Syst. 53 (3), 247–259.

Monsi, T.M., Saeki, T., 1953. Über den Lichtfaktor in den Pflanzengesellschaften und seine Bedeutung für die Stoffproduktion. Jap. J. Bot. 14, 22–52.

Morales-Ruiz, A., Loeza-Corte, J.M., Diaz-Lopez, E., Morales-Rosales, E.J., Franco-Mora, O., Mariezcurrena-Berasain, M.D., et al., 2016. Efficiency on the use of radiation and corn yield under three densities of sowing. Int. J. Agron. 5. Available from: https://doi.org/10.1155/2016/6959708. Article ID: 6959708.

Muchow, R.C., Robertson, M.J., Pengelly, B.C., 1993. Radiation use efficiency of soybean, mungbean, and cowpea under different environmental conditions. Field Crops Res. 32, 1–16. Available from: https://doi.org/10.1016/0378-4290(93)90017-H.

Munz, S., Graeff-Hönninger, S., Lizaso, J.I., Chen, Q., Claupein, W., 2014. Modeling light availability for a subordinate crop within a strip–intercropping system. Field Crops Res. 155, 77–89. Available from: https://doi.org/10.1016/j.fcr.2013.09.020.

Narayanan, S., Aiken, R.M., Prasad, P.V.V., Xin, Z., Yu, J., 2013. Water and radiation use efficiencies in sorghum. Agron. J. 105, 649–656. Available from: https://doi.org/10.2134/agronj2012.0377.

Neiff, N., Trachsel, S., Valentinuz, O.R., Balbi, C.N., Andrade, F.H., 2016. High temperatures around flowering in maize: effects on photosynthesis and grain yield in three genotypes. Crop Sci. 56, 2702–2712. Available from: https://doi.org/10.2135/cropsci2015.12.0755.

Nelson, K.A., 2014. Corn yield response to the solar corridor in Upstate Missouri. Agron. J. 106, 1847–1852. Available from: https://doi.org/10.2134/agronj2012.0326.

Ney, B., Bancal, M.O., Bancal, P., Bingham, I.J., Foulkes, J., Gouache, D., et al., 2013. Crop architecture and crop tolerance to fungal diseases and insect herbivory. Mechanisms to limit crop losses. Eur. J. Plant Pathol. 135, 561–580. Available from: https://doi.org/10.1007/s10658-012-0125-z.

Niinemets, Ü., 2010. A review of light interception in plant stands from leaf to canopy in different plant functional types and in species with varying shade tolerance. Ecol. Res. 25, 693–714. Available from: https://doi.org/10.1007/s11284-010-0712-4.

O'Connell, M.G., O'Leary, G.J., Whitfield, D.M., Connor, D.J., 2004. Interception of photosynthetically active radiation and radiation-use efficiency of wheat, field pea, mustard, in a semi-arid environment. Field Crops Res. 85, 111–124.

Pangga, I.B., Hanan, J., Chakraborty, S., 2013. Climate change impacts on plant canopy architecture: implications for pest and pathogen management. Eur. J. Plant Pathol. 135, 595–610.

Peleg, Z., Reguera, M., Tumimbang, E., Walia, H., Blumwald, E., 2011. Cytokinins-mediated source/sink modifications improve drought tolerance and increase grain yield in rice under water stress. Plant Biotechnol. J. 9, 747–758.

Purcell, L.C., Ball, R.A., Repaer III, J.D., Vories, E.D., 2002. Radiation use efficiency and biomass production in soybean at different plant population densities. Crop Sci. 42, 172–177.

Rachaputi, R.C.N., Chauhan, Y., Douglas, C., Martin, W., Krosch, S., Agius, P., et al., 2015. Physiological basis of yield variation in response to row spacing and plant density of mungbean grown in subtropical environments. Field Crops Res. 183, 14–22. Available from: https://doi.org/10.1016/j.fcr.2015.07.013.

Reguera, M., Peleg, Z., Abdel-Tawab, Y.M., Tumimbang, E.B., Delatorre, C.A., Blumwald, E., 2013. Stress-induced cytokinin synthesis increases drought tolerance through the coordinated regulation of carbon and nitrogen assimilation in rice. Plant Physiol. 163, 1609–1622.

Rochette, P., Desjardins, R.J., Pattey, E., Lessard, R., 1995. Crop net carbon dioxide exchange rate and radiation use efficiency in soybean. Agron. J. 87, 22–28.

Rosenthal, W.D., Gerik, T.J., Wade, L.J., 1993. Radiation-use efficiency among grain sorghum cultivars and plant densities. Agron. J. 85, 703–705.

Ross, J., 1975. Radiative transfer in plant communities. In: Monteith, J.L. (Ed.), Vegetation and the Atmosphere, vol. 1. Academic Press, London, UK, pp. 13–55.

Sellers, P.J., Berry, J.A., Collatz, G.J., Field, C.B., Hall, F.G., 1992. Canopy reflectance, photosynthesis, and transpiration. III. A reanalysis using improved leaf models and a new canopy integration scheme. Remote Sens. Environ. 42, 187–216.

Seran, T.H., Brintha, I., 2010. Review on maize based intercropping. J. Agron. 9, 135–145.

Sharkey, T.D., Bernacchi, C.J., Farquhar, G.D., Singsaas, E.L., 2007. Fitting photosynthetic carbon dioxide response curves for C_3 leaves. Plant Cell Environ. 30 (9), 1035–1040.

Sinclair, T.R., Horie, T., 1989. Leaf nitrogen, photosynthesis, and crop radiation use efficiency: a review. Crop Sci. 29 (1), 90–98.

Sinclair, T.R., Muchow, R.C., 1999. Radiation use efficiency. Adv. Agron. 65, 215–265. Available from: https://doi.org/10.1016/S0065-2113(08)60914-1.

Sinclair, T.R., Shiraiwa, T., Hammer, G.L., 1992. Variation in crop radiation-use efficiency with increased diffuse radiation. Crop Sci. 32, 1281–1284.

Singer, J.W., Sauer, T.J., Blaser, B.C., Meek, D.W., 2007. Radiation use efficiency in dual winter cereal-forage production system. Agron. J. 99, 1175–1179. Available from: https://doi.org/10.2134/agronj2007.0033.

Singer, J.W., Meek, D.W., Sauer, T.J., Prueger, J.H., Hatfield, J.L., 2011. Variability of light interception and radiation in maize and soybeans. Field Crops Res. 121, 147–152. Available from: https://doi.org/10.1016/j.fcr.2010.12.007.

Stewart, D.W., Costa, C., Dwyer, L.M., Smith, D.L., Hamilton, R.I., Ma, B.L., 2003. Canopy structure, light interception, and photosynthesis in maize. Agron. J. 95, 1465–1474.

Stirling, C.M., Aguilera, C., Baker, N.R., Long, S.P., 1994. Changes in the photosynthetic light response curve during leaf development of field grown maize with implications for modeling canopy photosynthesis. Photosynth. Res. 42, 217–225.

Stoy, P.C., Trowbridge, A.M., Bauerle, W.L., 2014. Controls on seasonal patterns of maximum ecosystem carbon uptake and canopy-scale photosynthetic light response: contributions from both temperature and photoperiod. Photosynth. Res. 119, 49–64. Available from: https://doi.org/10.1007/s11120-013-9799-0.

Tao, Z., Wang, D., Ma, S., Yang, T., Zhao, G., Chang, X., 2018. Light interception and radiation use efficiency response to tridimensional uniform sowing in winter wheat. J. Integrative Agric. 17, 566–578. Available from: https://doi.org/10.1016/S2095-3119(17)61715-5.

Tesfaye, K., Walker, S., Tsubo, M., 2006. Radiation interception and radiation use efficiency of three grain legumes under water deficit conditions in a semi-arid environment. Eur. J. Agron. 25, 60–70. Available from: https://doi.org/10.1016/j.eja.2006.04.014.

Tholen, D., Boom, C., Zhu, X., 2012. Opinion: prospects for improving photosynthesis by altering leaf anatomy. Plant Sci. 197, 92–101. Available from: https://doi.org/10.1016/j.plantsci.2012.09.005.

Thom, A.S., 1975. Momentum, mass, and heat exchange of plant canopies. In: Monteith, J.L. (Ed.), Vegetation and the Atmosphere, vol. 1. Academic Press, London, UK, pp. 57–109.

Thomas, H., Ougham, H., 2014. The stay-green trait. J. Exp. Bot. 65, 3889–3900.

Tian, G., Gao, L., Kong, Y., Hu, X., Xie, K., Zhang, R., et al., 2017. Improving rice population productivity by reducing nitrogen rate and increasing plant density. PLoS One 12, e0182310. Available from: https://doi.org/10.1371/journal.pone.0182310.
Timlin, D.J., Fleisher, D.H., Kemanian, A.R., Reddy, V.R., 2014. Plant density and leaf area index effects on the distribution of light transmittance to the soil surface in maize. Agron. J. 106, 1828–1837.
Tollenaar, M., Fridgen, J., Tyagi, P., Stackhouse Jr, P.W., Kumudini, S., 2017. The contribution of solar brightening to the US maize yield trend. Nat. Clim. Change 7, 275.
Torres, G.M., Koller, A., Taylor, R., Raun, W.R., 2017. Seed-oriented planting improves light interception, radiation use efficiency and grain yield of maize (*Zea mays* L.). Exp. Agric. 53 (2), 210–225.
Tournebize, R., Sinoquet, H., 1995. Light interception and partitioning in a shrub/grass mixture. Agric. For. Meteorol. 72, 272–294.
Vanaja, M., Sathish, P., Kumar, G.V., Razzaq, A., Vagheera, P., Lakshmi, N.J., et al., 2017. Elevated temperature and moisture deficit stress impact on phenology, physiology and yield responses of hybrid maize. J. Agrometorol. 19, 295–300.
Vance, R.R., Nevai, A.L., 2007. Plant population growth and competition in a light gradient: a mathematical model of canopy partitioning. J. Theor. Biol. 245, 210–219. Available from: https://doi.org/10.1016/j.jtbi.2006.10.015.
Verdelli, D., Acciaresi, H.A., Leguizamón, E.S., 2012. Corn and soybeans in a strip intercropping system: crop growth rates, radiation interception, and grain yield components. Int. J. Agron. 17. Available from: https://doi.org/10.1155/2012/980284. Article ID: 980284.
Von Caemmerer, S., Furbank, R.T., 1999. Modeling C_4 photosynthesis. In: Monson, R. K., Sage, R.F. (Eds.), C_4 Plant Biology. Academic Press, Toronto, Canada, pp. 173–211.
Wall, G.W., Kanemasu, E.T., 1990. Carbon dioxide exchange rate in wheat canopies. Part I. Influence of canopy geometry on trends in leaf area index, light interception and instantaneous exchange rates. Agric. For. Meteorol. 49, 81–102.
Wang, Q., Niu, Y., Xu, Q., Wang, Z., Zhang, X., 1995. Relationship between plant type and canopy apparent photosynthesis in maize (*Zea mays* L.). Biol. Plant. 37, 85–91.
Wang, X., Guo, Y., Li, B., Wang, X., Ma, Y., 2006. Evaluating a three dimensional model of diffuse photosynthetically active radiation in maize canopies. Int. J. Biometeorol. 50, 349–357. Available from: https://doi.org/10.1007/s00484-006-0032-0.
Wang, Y., Song, Q., Jaiswal, D., de Souza, A.P., Long, S.P., Zhu, X., 2017. Development of a three-dimensional ray-tracing model of sugarcane canopy photosynthesis and its application in assessing impacts of varied row spacing. Bioenergy Res. 10, 626–634. Available from: https://doi.org/10.1007/s12155-017-9823-x.
Warren Wilson, J., 1965. Stand structure and light penetration. I. Analysis by point quadrats. J. Appl. Ecol. 2, 383–390.
Warren Wilson, J., 1967a. Stand structure and light penetration. III. Sunlit foliage area. J. Appl. Ecol. 4, 159–165.
Warren-Wilson, J.W., 1967b. Ecological data on dry-matter production by plants and plant communities. In: Bradley, E.F., Denmead, O.T. (Eds.), The Collection and Processing of Field Data. Interscience Publishers, New York, pp. 77–123.
Westgate, M.E., Otegui, M.E., Andrade, F.H., 2004. Physiology of the corn plant. In: Smith, C., Betrán, J., Runge, E. (Eds.), Corn: Origin History, Technology, and Production. John Wiley & Sons, Hoboken, NJ, p. 282.

Yamori, W., Hikosaka, K., Way, D.A., 2014. Temperature response of photosynthesis in C_3, C_4, and CAM plants: temperature acclimation and temperature adaptation. Photosynth. Res. 119 (1), 101–117.

Zhan, D., Zhang, C., Yang, Y., Luo, H., Zhang, Y., Zhang, W., 2015. Water deficit alters cotton canopy structure and increases photosynthesis in the mid-canopy layer. Agron. J. 107, 1947–1957. Available from: https://doi.org/10.2134/agronj14.0426.

Zhao, J., Yang, X., Lin, X., Sassenrath, G.F., Dai, S., Lv, S., et al., 2015. Radiation interception and use efficiency contributes to higher yields of newer maize hybrids in Northeast China. Agron. J. 107, 1473–1480.

Zhu, X.G., Long, S.P., Ort, D.R., 2008. What is the maximum efficiency with which photosynthesis can convert solar energy into biomass? Curr. Opin. Biotechnol. 19, 153–159.

Zhu, X., Song, Q., Ort, D.R., 2012. Elements of a dynamic systems model of canopy photosynthesis. Curr. Opin. Plant Biol. 15, 237–244. Available from: https://doi.org/10.1016/j.pbi.2012.01.010.

Further reading

Kiniry, J.R., Tischler, C.R., Van Esbroeck, G.A., 1999. Radiation use efficiency and leaf CO_2 exchange for diverse C4 grasses. Biomass Bioenergy 17, 95–112. Available from: https://doi.org/10.1016/S0961-9534(99)00036-7.

Liu, X., Rahman, T., Song, C., Su, B., Yang, F., Yong, T., et al., 2017. Changes in light environment, morphology, growth and yield of soybean in maize-soybean intercropping systems. Field Crops Res. 200, 38–46. Available from: https://doi.org/10.1016/j.fcr.2016.10.003.

CHAPTER 2

Variety selection for the solar corridor crop system (SCCS)

C. LeRoy Deichman
Deichman Consulting, Shelbyville, MO, United States

Contents

Introduction

The initial objective of the solar corridor investigative inquiry was to enable increased corn yield beyond that which current best management practices (BMPs) deliver, with a projected exponential increase in net operating profits (NOPs) per corn acre and subsequent improvements in soil health and sustainability of the farm unit.

The Solar Corridor Crop System
DOI: https://doi.org/10.1016/B978-0-12-814792-4.00002-4

Procedure

The preliminary step of our inquiry was to identify hybrids that were likely to be sufficiently responsive to the enriched solar corridor production environment. That step, outlined later in this chapter, identified 21 (of 200) high-yielding non-Genetically Modified Organism (GMO) hybrids as likely candidates to attain optimized performance in properly designed, developed, and deployed solar corridor production systems. Indeed, success of the system itself depended on detecting which corn hybrids would, in fact, be most responsive to the drastically different corn production paradigm, or potentially enhanced corn production environment, offered by the solar corridor.

The setting: the solar corridor and its enabling factor

The solar corridor complements and builds on previously accomplished productivity using the development and management of a finite soil base and often finite water resources by a more efficient utilization (incident sunlight) and more effective capture (atmospheric carbon dioxide (CO_2)) of the relatively infinite atmospheric resources. The enabling factor focuses on how the plant's photosynthetic organs receive the premier photochemical energy source of incident sunlight. That focus enables the plant to capture more atmospheric-sourced CO_2, which allows the plant to make more efficient use of all other essential plant production inputs. It is critically important that the comprehensive basis of this plant production principle is understood. Von Liebig's law of the minimum is most instructive (and other chapters of this book including "Putting It All Together"). Furthermore, effective plant capture of infinite quantities of atmospheric-sourced nitrogen and more finite precipitation is made possible by many solar corridor crop system (SCCS)—enabled crop options.

General background

While this book is intended to distill and summarize the current status of the scientifically conducted and validated research on the solar corridor concept and biological basis of the solar corridor, this chapter focuses specifically on (corn) variety selection and its vital role in maximizing the performance that consistently has been, and can be, achieved in the properly designed, developed, and deployed SCCS wherever corn is chosen as the "main" crop. In its solar corridor context, the "main" crop is the principal (crop) beneficiary of the incident sunlight within the enriched solar corridor production environment. In contrast, the "floor" crop is the crop that occupies the floor, or interrow space, of the solar corridor while the main crop maximizes dominance in capturing incident sunlight—the premier photochemical energy source for photosynthesis—which (exclusive of the corn grower) is the largest contributor, but not the largest consumed component (CO_2), of corn yield.

To the best of my knowledge, only four research entities have conducted significant research on variety selection for the solar corridor. Although one ongoing project (Reinbott et al., 2018) is not yet published and one project is proposed for 2019, we shall proceed with an overview of all of the currently published results because the information is critical for demonstrating the vital role of variety selection for successful corn production in the hybrid-specific SCCS. Twenty-one of the first 200 high-yielding commercial corn hybrids that we screened in our preliminary hybrid screening trial (PST) met the criteria that we hypothesized would indicate the potential to enable successful performance in the SCCS. Three of those hybrids, designated as hybrids B, C, and D were selected for our advanced seminal research studies. Another, designated as hybrid A, was one of the 179 that failed to meet the above criteria.

Though a student of interdisciplinary corn production nearly all of my life, I know only a fraction of what I "would like to know" about corn production. But if you have a stake in corn production, I would like you to know what I "do know" about a few basic facts or benchmark conclusions that, at first, may seem inconceivable.

Fact 1

With properly chosen high-yielding commercial corn hybrids using state-of-the-art selection criteria and state-of-the-art BMPs, the SCCS resulted in yield levels considerably above US corn grain trend line yields (Nielsen, 2017) from seven different soils at 40–41°N latitude in the heart of the US Corn Belt, and on the predominant prairie soils of NE Missouri. The solar corridor enabled the production of as much or more corn from one 152-cm-wide twin corn row as was produced from two state-of-the-art BMP 76-cm (30-in.) corn rows. The soils at the field sites in the Corn Belt region were Bryce silty clay/Swygert silty clay loam (fine-silty, mixed, superactive, mesic Aquic Argiudolls/fine, mixed, active, mesic Aquic Argiudolls); Elliott silt loam/Ashkum silty clay loam (fine, illitic, mesic Aquic Argiudolls/fine, mixed, superactive, mesic Typic Endoaquolls); Milford (fine, mixed, superactive, mesic Typic Endoaquolls); Brenton silt loam (fine-silty, mixed, superactive, mesic Aquic Argiudolls); and Gilford sandy loam (coarse-loamy, mixed, superactive, mesic Typic Endoaquolls); soils in northeast Missouri were Mexico silt loam (fine, smectitic, mesic Vertic Epiaqualfs) and Putnam silt loam (fine, smectitic, mesic Vertic Albaqualf). The description and results of the seminal studies conducted in the Corn Belt sites are found in Deichman and Nelson (2000), Deichman et al. (2001), and Deichman (2002). Also, yields at the Bryce and Swygert soil sites, while mirroring the summary yields, were only used 1 year, and subsequently not included in summaries reported in the scientific literature.

Standard testing protocols were developed based on the seminal screening studies conducted during the 1990s (Fig. 2.1). This was necessary to demonstrate the yield performance of solar corridor treatments compared to state-of-the-art conventional BMP control treatments. The PST consistently showed significant differences due to hybrid response to the solar corridor treatments, thus testing in the SCCS needs to be split by hybrid to capture the actual hybrid-specific response. Hybrid performance also differed highly by population density, indicating that another split treatment by plant population study protocol was designed to obtain a more comprehensive and definitive response to the solar corridor environment.

Using four hybrid entries in the seminal studies provided the opportunity to observe three basic phenotypic characteristics that influenced

Treatment protocols

- 12 production environments
- 4 hybrids (designated A, B, C, and D)
- 4 plant populations (1/2X, 2/3X, 5/6X, and X)
- 3 replications
- Randomized block split/split plot design
  - 1st split by hybrid, 2nd by plant population
- 2 row width entries
  - Control: Single rows on 76 or 91 centimeter centers
  - Treatment: Twin rows on 152 or 183 centimeter centers
- All treatments were in north/south rows between 40 and 41°N latitude

Figure 2.1 Treatment protocols used in the trials with results shown in Figs. 2.2–2.4. *Deichman Consulting and Maize Research Unlimited, Inc.*

potential yield increases in the SCCS by three selected hybrids designated B, C, and D relative to a high-yielding modern commercial hybrid (A) that failed to respond sufficiently to the SCCS. We found 21 commercial hybrids out of 179 high-yielding hybrids that met the PST response criteria when screened using the protocol for the advanced seminal studies (Fig. 2.1). The PST protocol is outlined later in this chapter. The most successful state-of-the-art BMP controls undergoing PST had a predetermined plant population of 30,000 plants acre^{-1} (74,100 ha^{-1}) while the solar corridor treatments were 15,000 plants acre^{-1} (37,000 ha^{-1}) thereby establishing the wide (X) to one-half X population ranges used in the protocol. This also gave us the opportunity to see if the hybrids identified by the PST would perform as predicted over a wide range of conditions prevalent in the area where the PST were conducted.

This following summary data highlights the important findings resulting from numerous studies conducted during the 1990s (until the hybrids used in the studies were no longer available). These summary studies are based on individual split–split plots with hybrid and plant population treatment factors replicated three times (Fig. 2.1). Individual grain yields ranged from 100 to 228 bu acre^{-1} (6200–14,300 kg ha^{-1}) (Table 2.1). Note that the highest yield (228 bu acre^{-1} (14,300 kg ha^{-1})) is reported in yield tables of Deichman (2000). Results showed that the three selected hybrids grown in eastern Illinois at 40–41°N latitude produced as much or more grain yield in 152-cm twin rows as in 76-cm conventional row widths and, similarly, 183-cm twin row planting produced as much or more grain yield as 91-cm row widths. This response was consistent, thus average yields for 12 environments and four plant populations demonstrate the yield advantage obtained with the solar corridor (Fig. 2.2). Highest treatment yields were obtained from the highest plant populations from hybrids C, B, and D; the lowest were of the lowest populations of A, B, and C. Suboptimal low populations for hybrids A through C give a more expansive range of conditions than we would expect to encounter in typical cornfields in this area of eastern Illinois at 40–41°N latitude. This supports the ability of the PST to identify hybrids for the solar corridor enhanced production environment, but the yield summaries with all populations included are rather low and we do not assume that response will be the same across all yield ranges. Thus Fig. 2.3 shows the average yields across the eight highest performance environments, at all populations including the three suboptimal that may result in different responses. Thus, the double check, confirming that the positive response of B, C,

Table 2.1 Solar corridor crop system (SCCS) trial data and results.

	Study								Control	
	Location	Latitude	Varieties	Row orientation	Row spacing	Population	Soils	Yield (bu acre^{-1})	Row spacing	Yield (bu acre^{-1})
1[a]	E Central IL and S Central MN	40° 17′–44° 43′ N	20 corn	0–180° to 90°–270°	56–76 in.	15,000–33,000	Various	48–228.3	28–38 in.	29[b]–226.9
2[c]	E Central IL	40° 28′, 40° 30′, and 40° 46′ N	DEKALB 711, Cargill 7877 and Prairie Stream 702	0–180°	60 and 72 in.	30,000	Brenton, Elliott, Ashkum and Gilford Very Fine Sandy Loam	213	30 and 36 in.	193
3	E Central IL	40° 28′ N	Prairie Stream 702	0–180°	72 in.	30,000	Brenton silt loam	217	36 in.	192
4	E Central IL	40° 28′ N	Prairie Stream 702	0–180°	60 in.	30,000	Elliott silt loam (North 4), Ashkum silty clay loam (South 4)	211	30 in.	178
5	E Central IL	40° 28′, 40° 30′, and 40° 46′ N	Prairie Stream 702	0–180°	60 and 72 in.	30,000	Brenton, Elliott, Ashkum and Gilford Very Fine Sandy Loam	214	30 and 36 in.	174

6	E Central IL	40° 46′ N	Prairie Stream 702	0–180°	60 in.	30,000	Gilford Very Fine Sandy Loam	213	30 in.	153
7	E Central IL	40° 46′ N	Cargill 7877	0–180°	60 in.	30,000	Gilford Very Fine Sandy Loam	219	30 in.	214
8	E Central IL	40° 30′ N	Cargill 7877	0–180°	60 in.	30,000	Elliott (North 4), Ashkum (South 4)	228	30 in.	214
9	E Central IL	40° 28′ N	Cargill 7877	0–180°	72 in.	30,000	Brenton	227	36 in.	207
10	E Central IL	40° 46′ N	DEKALB 711	0–180°	60 in.	30,000	Gilford Very Fine Sandy Loam	192	30 in.	164
11	E Central IL	40° 28′ N	DEKALB 711	0–180°	72 in.	30,000	Brenton	221	36 in.	227

[a]Average for all SCCS trials using 20 different corn varieties.
[b]Low yield fields were subject to adverse conditions, such as drought.
[c]Average for experiments using only three of the seven commercially available corn varieties as identified in the table.
Source: Information based on Deichman, C.L., 2000. Plant arrangement for improving crop yields. U.S. Patent 6,052,941. Date issued: 25 April 25, 2000.

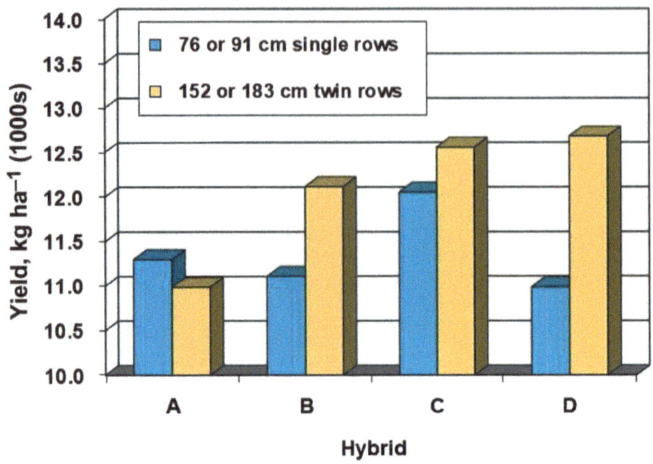

Figure 2.2 Response of hybrids to row spacing, showing average yields across all 12 production environments, including four plant populations. Hybrids B and D produced significantly higher yields for the widely spaced twin rows. Hybrid A performs worse. *Deichman Consulting and Maize Research Unlimited, Inc.*

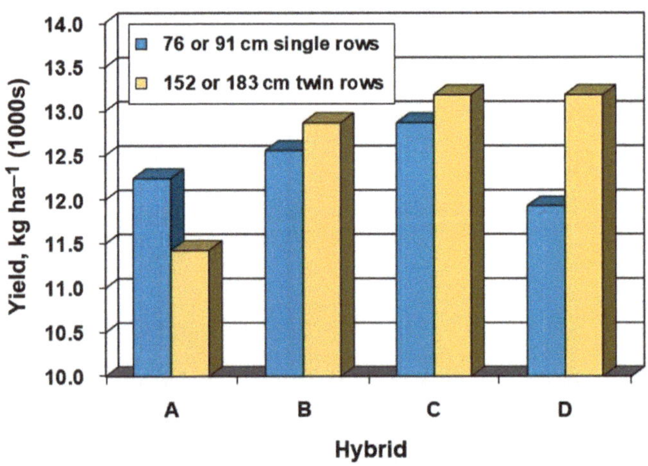

Figure 2.3 Response of hybrids to row spacing, showing average yields across the eight highest yielding environments, including four plant populations. *Deichman Consulting and Maize Research Unlimited, Inc.*

and D is retained throughout the yield and population ranges that our studies evaluated. Fig. 2.4 shows the yield summaries of all locations, without the three lowest plant populations. Likewise, we don't assume the same responses through all populations. Again, the positive responses of hybrids B, C, and D to the solar corridor production environment are retained but at different response magnitudes.

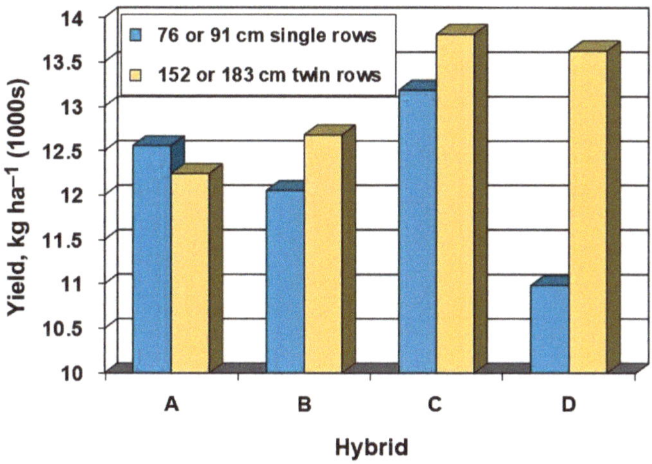

Figure 2.4 Response of hybrids to row spacing, showing yields at the highest plant population. *Deichman Consulting and Maize Research Unlimited, Inc.*

We found that each hybrid may perform best at slightly different population thresholds, the general population response can be illustrated by hybrid B on the Brenton silt loam site (Figs. 2.5 and 2.6), which is representative of the response trend of the three hybrids (A, B, and C). However, an exception to the trend is the response of hybrid D at each test site, shown in Fig. 2.6. Furthermore, hybrid D planted in 72-in. (183-cm) wide rows on a Gilford fine sandy loam site, produced over 200 bu $acre^{-1}$ (12,540 kg ha^{-1}) at all populations (15,000 through 30,000 plants $acre^{-1}$), which were up to 60 bu $acre^{-1}$ (1400 kg ha^{-1}) more than it produced at the same population in state-of-the-art BMP controls (Table 2.1 and see Patent Table 9 in Deichman, 2000).

The above trial results validate the potential yield response of hybrids selected for production in the SCCS as disclosed in the patent issued for "a planting arrangement to improve crop yield" (Deichman, 2000). The yield data and results required as evidence to support patent claims are summarized in Table 2.1 but do not include four hybrids that failed to achieve PST performance criteria. The patent (Deichman, 2000) does not specifically disclose how to select specific hybrids required to maximize site-specific SCCS yields. The PST protocol we used (described as the specific conservative method (CM) methodology later in this chapter), demonstrated by the yield summaries (Figs. 2.2–2.6), patent data Tables 2.1–2.14 (Deichman, 2000) summarized in Table 2.1, and full details in the data analysis (https://www.agronomy.org/files/about-agronomy/data-from-1992-sccs-trials.xlsx), was

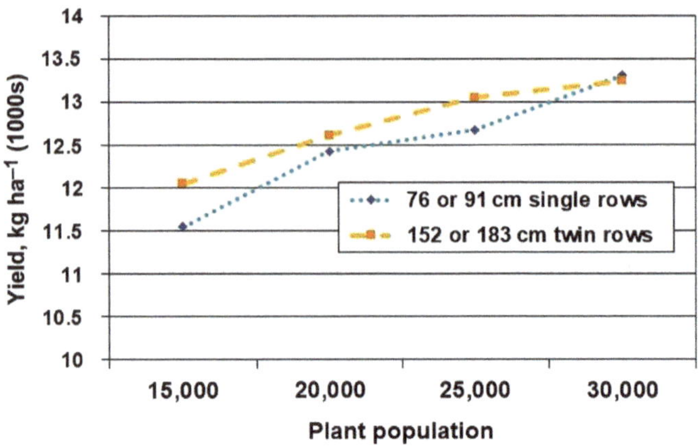

Figure 2.5 Plant population and row width: Brenton silt loam—hybrid B example. *Deichman Consulting and Maize Research Unlimited, Inc.*

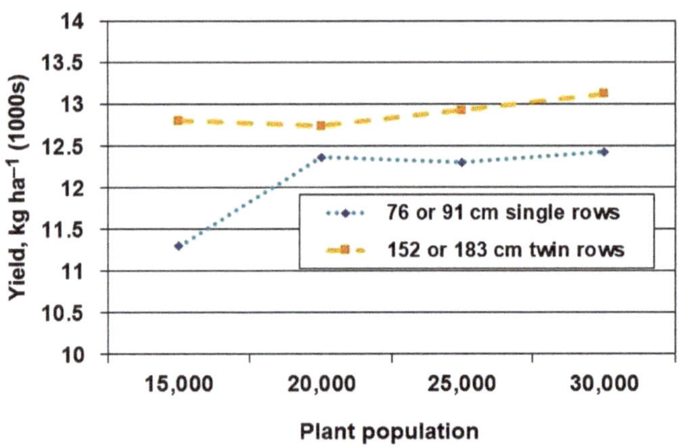

Figure 2.6 Plant population and row width: Brenton silt loam—hybrid D example. *Deichman Consulting and Maize Research Unlimited, Inc.*

very effective in identifying high-performing hybrids for the solar corridor across a wider range of production environments than the US Corn Belt producer would typically encounter.

We hypothesize that the source of and basis for increased productivity of corn grown in the SCCS is the efficient capture of incident solar radiation by a leaf canopy fully exposed to sunlight through photosynthesis and carbon-based photosynthate compounds that subsequently translocate into and are transformed through the plant metabolic pathways that contribute to improved grain yields, as illustrated in Fig. 2.7. Thus as our yield data

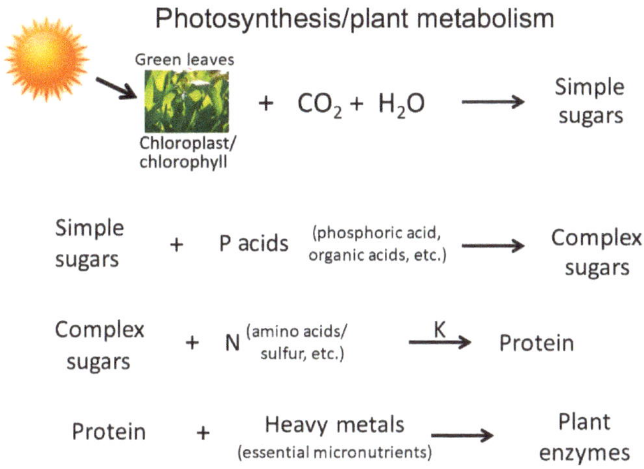

Figure 2.7 In this summarized context, enzymes link the complex genetic by environment interaction where a unique phenotypic response is expressed in crop yield.

indicate, hybrid selection, as governed by the following principles, is an integral component of resulting crop yield. These principles are as follows:

1. $R \times D = Y$, rate and duration of photosynthesis, and subsequent biological transformations delivered to meet reproductive sink demand gives the phenotypic expression of quantitative crop yield;
2. Von Liebig's law of the minimum may be illustrated by the stave barrel, that is, the shortest stave determines its capacity as the most limiting factor imposes the effective ceiling on crop yield. For an illustration and discussion of von Liebig's law of the minimum, see Brady and Weil (2008, Chapter 16);
3. The yield value of Input A is a function of the support given by Inputs B through Z (while there are more inputs than letters of the alphabet, this illustrates the resulting synergies);
4. Yield, as supported or limited by all (and specific) environmental stimuli, is expressed by the quantitative genetics of the sourcing crop. Or . . .

Except as limited by specific input or environmental event, yield is determined by phenotypic utilization of the aggregate but time-sensitive production environment.

Fact 2

On the other half of the cornfield, consisting of the interrow area between the wide corn rows which form the base, or floor, of the solar corridor, we can produce a normal yield of specific, high-value rotational floor crops, that is compatible and without negative impact on corn yield, effectively doubling corn yield per acre (Deichman, 2007; Deichman et al., 2016). An example of the solar corridor model, presented previously (Deichman, 2009), is shown in Fig. 2.8, which illustrates corridors between 60-in. (152 cm) twin rows of corn that enable incidental sunlight to reach each leaf from the early vegetative through reproductive growth stages; the solar corridor floor can be planted to any number of "low-stature" crops such as winter wheat with red clover, spring oats, field peas, cowpea, or other cover crops, which may continue to grow after corn harvest in autumn in temperate climate regions.

Abundant, accessible, basic, low-cost, and sustainable God-given sunshine, redeployed through the SCCS that enables the production of as much additional corn yield as the current and laudable 79-year average trend line yield increases (Nielsen, 2017) and as much as all previous years of corn production, *combined*, have produced, is, understandably "hard to get your arms around." Accordingly, our readers should have every

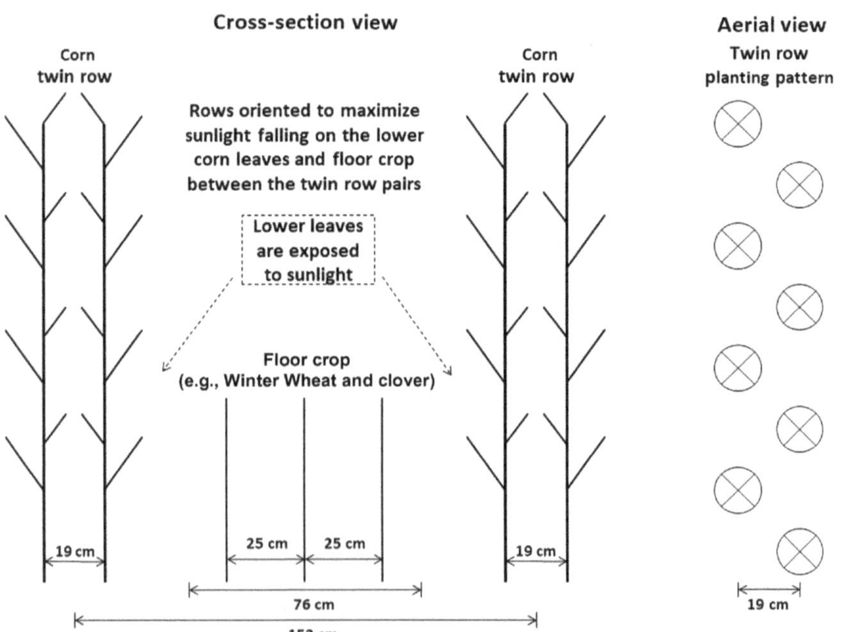

Figure 2.8 Solar corridor crop system design. Modified from Deichman, C.L., 2000. Plant arrangement for improving crop yields. U.S. Patent 6,052,941. Date issued April 25, 2000. Diagram by H.I. Britt.

opportunity to fully examine the entire basis of such inconceivable yield results. Therefore a link (https://www.agronomy.org/files/about-agronomy/data-from-1992-sccs-trials.xlsx) to the statistical analysis spreadsheet is provided that offers a summary of the extensive yield data and statistical analysis that otherwise could not be presented within the limited confines of one 200-page book. It offers an option for your further study of the individual treatments, treatment combinations, and replications as you may choose, to help comprehend the unprecedented basis of the benchmark conclusions of Facts 1 and 2.

With arable land as our most finite crop production resource, the focus of the SCCS on increased productivity per (spatial) arable land unit rather than on productivity per worker unit, offers an actual and significant quantity of increased sustainable production that is beyond the scope of current crop production paradigms. By this point, we have hopefully answered to your satisfaction the yield portion of the initial objective of the solar corridor as stated at the beginning of this chapter and documented that *variety selection* is basic for maximized performance of the SCCS.

Fact 3

Since the extra yield sourcing inputs (sunlight and CO_2), related to Facts 1 and 2, are abundant, relatively inexpensive (no direct charge), and sustainable, personal deployment of the extra human capital required by the SCCS enables extra NOP per spatial land unit (acre, hectare, etc.). Undoubtedly, NOP is of more personal value than yield alone.

Fact 4

The SCCS enables the biosynthesis of more atmospheric CO_2 and more carbon substances exuded through corn roots and transferred into the rhizosphere soil to enrich the labile and active carbon pool (Paterson, 2003). These processes are critical for sustaining the soil and rhizosphere microbial community, which in turn provide crop growth promoting substances and contribute to formation of vital soil organic matter and to more available soil moisture holding capacity (Kremer and Deichman, 2014; Kremer et al., 2015; Schmidt et al., 2016).

Fact 5

The properly designed, developed, and deployed SCCS also reduces soil erosion, soil nutrient loss, and soil compaction and increases the effective capture of precipitation at the point of soil impact thereby further increasing available soil moisture storage. Further discussion on soil plant–water relationships relative to the SCCS is provided in Chapter 3, Water-use efficiency in the SCCS, of this volume (Timlin et al., 2019).

Fact 6

Anecdotal evidence shows that the solar corridor has demonstrated a nearly inconceivable resistance to wind and stalk rots of multiple pathogens in production fields on both sides of the Mississippi in Marion County, MO, and Ford County, IL (Alex Carpenter; Danny and Gary Bevill; Darrell Smith; Rick and Stan Nelson; C.L. Deichman, personal observations, 2000 and 2010). We can hypothesize that the reproductive sinks do not capture all of the extra carbon-based photosynthetic compounds. Thus, vegetative (root and stalk) sinks benefit also (e.g., nearly every plant of hybrid D would lodge before ears began to dent under high populations in state-of-the-art BMP cornfields). It exhibited unbelievable stalk strength in the enriched solar corridor production environment!

Historical background and development of the solar corridor crop system

From the dawn of civilized agriculture, the most basic limiting factors of crop production have been resolved, that is, water from the Nile River provided water for the crops on parched land that didn't have enough water, which enabled them to become productive. Some of today's most productive soils had to be drained for the plant roots to obtain the required amount of oxygen needed before a crop of corn could be produced. Thus, irrigation and drainage successfully address two of the most basic yield-limiting factors: root available water (H_2O) and oxygen (O_2).

More recently, the dawn of modern agriculture was ushered in by the discoveries of hybrid corn and synthetic nitrogen and quickly (about 30 years) followed by robust seed and fertilizer industries including the current 360 degree complement of crop production support services that enable delivery of various supporting practices. After the extensive, exhaustive, and successful efforts of removing the myriad of currently identified and recognized site-specific (some of which are referenced above) yield-limiting factors, we propose the SCCS as our means of removing the next most limiting factor—incident sunlight, deep in the lower canopy of the cornfield throughout its critical reproductive stage through physiological maturity.

The chronology of the SCCS began with the earliest contemplation that it must be variety (hybrid) specific for it to function properly. This has been documented in a sequence of studies reporting results of periodic field trials that led to the creation of the SCCS (Deichman and Murrell, 1977; Deichman and Barrett, 1978; Deichman, 1980; Deichman and Warsaw, 1988; Deichman, 2000; Deichman and Nelson, 2000; Nelson, 2014; Reinbott et al., 2018; Smith, 2000). I conducted 60-in. twin row studies with a commercial-sized planter (Kenze 8-row, twin 30-in.) in a 40-acre field (Champaign County Illinois) in 1983 when I still had access to the genetics disclosed in early reports (Deichman and Murrell, 1977; Deichman and Barrett, 1978; Deichman, 1980), which illustrated how hybrids could be effectively segregated [e.g., by seven plant nutrients at Bayes Least Signifcant Difference (BLSD) 0.01] for different "well-defined" production environments. I hypothesized that, with full disclosure of the genetics in the hybrids—with properly sourced and sufficient soil and plant analysis data—we should be able to effectively select the hybrids for the solar corridor.[1] It is, however, a gross understatement to say that I was not incentivized to disclose proprietary information on the male and female parents in the varieties designated as commercial hybrids by the retail seed industry. When the seed company (CFS hybrids) whose formation was inspired by the results of my 1968 interdisciplinary studies in the northern Corn Belt, Lance Murrell's in the eastern Corn Belt, and other participating agronomists (Barrett, Gohlke, Darst, Wehunt, Wyatt et al.), which split soil fertility treatments by hybrid, was purchased by one of the largest seed companies in the world, my vision of the effective, simple selection of commercial hybrids for the solar corridor encountered a "setback." Having worked as an Interdisciplinary Corn Production Agronomist on the CFS corn breeding team in Minnesota and in various systemwide capacities for many years, I fully understood the seed company's interest or mandate to deny indiscriminate disclosure[2] of the genetics in their hybrids, but I disagreed with how their industrywide mandate was conducted. I contended that it helps no one but the seed company,

[1] The PST should not be equated to many subsequently heralded "genetic by environment" proprietary seed marketing terminologies that are based on less "well-defined" production environments.

[2] Disclosure of seed company's hybrid seed genetics presented by Deichman and Barrett (1978) and Deichman (1980).

who is often disingenuous about significant proprietary details of its hybrid. That disconnect deprives us of maximizing interdisciplinary corn production decisions that could help discerning corn growers. One of the pragmatic reasons this segment is included in this book is, in the event any seed company that prioritizes the interests of its customers chooses to reach out to us, we should then have the potential to expeditiously and seamlessly select and develop hybrids most responsive to the enhanced site-specific production environment offered by the solar corridor, but not necessarily limited to the solar corridor.

Thus pending an effective arrangement for one of the two proposals mentioned in this chapter, unless you know a seed supplier where hybrids are screened through an effective population (wheel) study on soils and production environments identical to those on your farm, or you have a uniquely extensive detailed data set on your site-specific production environment and hybrids,[3] a preliminary screening trial (PST) of corn hybrids should be conducted to determine which hybrid or hybrids to use in your site-specific production environment. As evidenced by this chapter, our experience says we should recognize and accept the caution of our conclusion that most high-yielding corn hybrids typically give a negative response to the solar corridor.

To conduct a PST we suggest telling your chosen seed supplier that you want small samples of a dozen or more (the more you evaluate the more likely you are to discover the hybrids that maximize solar corridor performance under your production environment) hybrids that exhibit:

1. exceptionally aggressive reproductive sinks, (strong sink pull);
2. an extended duration of chloroplast productivity or peak productivity; and
3. production in seed fields this year for planting next year.

As you conduct your PST as outlined below, we further suggest that you offer an invitation to agriculture extension agents, education entities (4-H/FFA), or research personnel of crop input or service providers in your community to encourage participation in hybrid performance evaluations.

[3] If an extensive detailed data set on your site-specific production environment and hybrids are available before initiating the SCCS, the farmer is still urged to confer with Deichman before completely relying on the information as one missing component (not apparent) could be the linchpin (a critical missing factor).

Protocols for preliminary corn hybrid screening trials

The easiest method

To use the easiest method (EM), on a relatively uniform site, using your BMPs and with a plant population near the high end of the population range (HP) recommended by the seed company for the hybrid, conduct the following:

1. Plant "control" rows as you normally plant, that is, every row planted;
2. For "treatment" rows, turn off seeding for every other row of the planter for the width of your harvest swath plus one row (so every harvest swath row is flanked by a blank row on both sides of the row);
3. Proceed planting for a distance required to obtain a row length adequate for an accurate yield measurement with the available equipment;
4. Replicate as deemed necessary for your specific situation;
5. Repeat the previous steps with each hybrid that meets the criteria above.

These planting trials may be in the form of randomized strip plots or randomized blocks where each hybrid entry in the block is either treatment or control. The next block would be the other treatment or control.

During the growing season, scout as necessary for effective pest control. For weed management, use effective weed control based on known weed problems for the field to combat the enriched plant growth environment. Insect pests that clip the last silks usually have a nonsignificant effect on yield; however, under the enriched production environment of the solar corridor, kernels that would never have developed on the end of the cob now have access to the additional photosynthate-sourced metabolites to develop. For example, in the previous solar corridor treatments, hybrid B kernel abortion on the end of the long ear was essentially eliminated. We also typically see a lower incidence of disease, and plant lodging is essentially eliminated via enhanced stalk strength, which I hypothesize to be due to a more robust carbohydrate supply in the lower plant canopy.

A quick stand appraisal should be taken after plant establishment to make sure your final control population is in the HP range. Treatment populations will have the same in-row plant spacing and half the plant population per field acre. If the control population is lower than HP, the yield results will be skewed by input contamination[4] and will be

[4] The production input we are measuring is extra incident sunlight facilitated by the solar corridor for penetration deep into the plant canopy. Wherever a plant is missing, extra incident sunlight is allowed to penetrate deeper into the plant canopy, around the missing plant, thus giving the "control" a portion of the production input that skews the overall measured impact of incident sunlight, that is, the measured response to the solar corridor is exponentially less responsive than it actually would be under a full population (HP). PST experiments where plant populations are substandard should be scrapped.

exponentially flawed. PST sites with plant populations lower than the higher ranges of the seed company recommended control population for the hybrid should be scrapped and not used in any yield summaries.

With verification and satisfaction of the previously mentioned protocol, with HP plant population as intended, using the same planter rows (easy to overlook in strip trials) for treatment and controls, with a border swath of at least six rows on each side of the control to prevent treatment contamination, measure, and compare yields of the "treatment" and "control" rows to obtain unique data showing how the hybrids included in your PST responded to your solar corridor production environment.

The conservative method

The CM was used to successfully identify hybrids B, C, and D in the seminal studies. On a relatively uniform site, using your BMPs and with a plant population near the HP recommended by the seed company for the hybrid, conduct the following:

1. Plant "control" rows as you normally plant, that is, every row planted;
2. As close as possible to the longest day of the year, chop down at least 30 ft. (at least 10 ft. longer than your intended harvest length) of corn on each side of the row you choose to be your solar corridor treatment row. Do a plant count and harvest the center 18–21 ft. of your solar corridor treatment row;
3. Lock that harvest length into your chain or measuring device and move it (at least 10 ft.) down the same row until you find the exact number of plants that you harvested in the "treatment." Weigh the corn from each;
4. Replicate as deemed necessary for your specific situation;
5. Repeat the previously mentioned steps with each hybrid that meets the criteria A, B, and C.

Upon completion you will have original specific information from your farm that hopefully allows you to discover a hybrid or two that responds to the solar corridor under your conditions. If you do not find (or until you do find) the hybrid(s) that meets our flex criteria, we withhold our recommendation that you adopt the solar corridor for your farm. If you decide to repeat the PST until you find the appropriate hybrid(s), include and repeat those that are most promising and evaluate with another set of hybrids, we most heartily support and agree with your decision.

Other preliminary corn hybrid screening trial research

In 2005 Nelson (2014) conducted a PST of six GMO hybrids using a randomized block design as described in EM above, which differs from seminal SCCS studies that were based on non-GMO hybrids B, C, and D using a more conservative design. The latest modern stacked GMO hybrids were used in this study, whereas the seminal studies used non-GMO hybrids introduced in the late 1980s following the CM method as described above to identify the likely responsive hybrids designated as hybrids B, C, and D in the seminal studies. His study showed that one hybrid (Burrus 794S) achieved both of the stringent criteria previously used to identify good candidates for the solar corridor: expression of the flex trait when grown at half of the high range of the recommended plant population (obtained by deleting every other row) to produce 90% of the control yield at the state-of-the-art BMP; and 85% of the yield of the highest yielding hybrid of an equivalent chronological development pattern in the PST before it is considered as a good candidate for the solar corridor. Determination of the physiological development stage for hybrid maturity qualification, as previously described in methods, was not necessary because Burrus 794S was the highest yielding hybrid in this study. It yielded 176 bu acre^{-1} at one-half of the recommended plant population. Another hybrid met the flex criterion, and the other four hybrids showed a typical negative response to the solar corridor. The response is typical based on compilation of a relatively limited database (Deichman, 2002; Deichman and Nelson, 2000; Deichman et al., 2001) indicating about 90% of the high-yielding commercial hybrids (179 of the first 200) assessed through the seminal PSTs failed to achieve the arbitrary criteria. North/south rows performed at a better than 10% achievement rate, while east/west rows reached a little less than 10%. Consistent with previous experience in Illinois, the 176 bu acre^{-1} Missouri yield was above 2005 trend line yields.

Unfortunately, the hybrid-specific results of the well-conducted 2005 Missouri study have not been published. Solar corridor studies subsequent to 2005 and 2006 (Nelson, 2014) did not include Burrus 794S because it was no longer commercially available. Table 1 in Nelson (2014) presents yields averaged over all hybrids planted at half the recommended plant population in the SCCS single-row configuration. This showed that the hybrids failed to yield similarly to the full population in conventional row widths because the SCCS planted at full population was not a study component. The yield potential of Burrus 794S at optimum population in the SCCS is unknown. Thus we can only speculate that the potential productivity of Burrus 794S would be consistent with our previously stated benchmark conclusion that "with properly chosen corn hybrids, the properly designed, developed and deployed SCCS produces as much or more corn from one 152–183 cm wide twin row as it produces from two state of the art 76–91 cm (30–36 in.) wide rows, respectively." On careful examination, Table 2 in Nelson (2014) did replicate the previous benchmark conclusion with another hybrid at 152-cm row width on prairie soil typical of northeast Missouri.

Further resolution of variety selection via patent applications pending submission, which provides a development technology for developing hybrids specifically for the SCCS, should significantly advance deployment of the SCCS. We conducted our hybrid PSTs to identify hybrids that offered the flex response potential to qualify for use in the advanced solar corridor studies. The advanced studies utilized a scientific protocol that took the solar corridor concept through the basic scientific process, which relies on others to conduct independent peer review thereby refining our original studies into seminal studies.

The hybrid PSTs produced some amazing and hard-to-believe results. Twenty-one (of the first 200) high-yielding commercial hybrids met both of the original response criteria, thus demonstrating a good potential to be sufficiently responsive to the SCCS production environment. We could fit only four hybrids into the advanced study protocol. For response comparisons, it was logical that one of the four should come from the 179 hybrids that did not meet the criteria. Accordingly, it was entered as hybrid A. The other three hybrids (B, C, and D) presented three distinctly different phenotypic expressions to maximize the quantitative output (yield) of the new light-enriched production environment provided by the solar corridor. Observable differences included more kernels (hybrid B), more ears (hybrid D), and a less distinct nature but a more balanced multiexpression (hybrid C), which contributed to heavier kernels. The first year of the advanced seminal studies, the drought year of 1991, produced unprecedented results (as much or more corn from one wide twin row as from two conventional state-of-the-art narrow single rows) that our new founding group, naturally, found hard to believe or explain except that it must have been due to the severely reduced yield of the 50-day summer drought (i.e., even though a nearly inconceivable response to the solar corridor—enriched production environment was obtained in a drouth year does not mean to expect same in a good year). That uniquely incentivized our team to determine the type of results that were possible in a good year. To our amazement, we produced the same results (as much or more corn from one wide twin row as from two conventional state-of-the-art narrow single rows), only at yields 100 bu $acre^{-1}$ higher. Our revised conclusion was that performance was not due to drought stress, but it could be affected by the small field plot arrangement. We tested that conclusion by using large field-scale strip plots and obtained the same 2X yield responses; thus use of small plots was not a factor. By this time the parallel study had been conducted showing that we could grow another crop on the other half of the (vacated) rows, or what we routinely call the "floor" crop without compromising corn yield (Crescent City 1 and 2; Deichman et al., 2016).

Conclusion

The solar corridor benefits as discussed in this book would not have occurred without the PST as addressed and described in this chapter. Our research based on the average of all state-of-the-art high-yielding hybrids released in the late 1980s indicates that the solar corridor configuration produced less than state-of-the-art BMP control production systems. However, non-GMO hybrids "selected" for the solar corridor production environment using the PST based on yields above current trend line yields (Nielsen, 2017) produced as much or more corn from one twin row as the state-of-the-art BMP controls produced from two rows. Therefore, any conclusion on solar corridor performance without it being hybrid specific does not accurately represent the SCCS. Accordingly, the PST is the logical first step for evaluating and maximizing solar corridor yield and subsequent performance objectives for your specific production environment.

References

Brady, N.C., Weil, R.R., 2008. The Nature and Properties of Soils. Pearson-Prentice Hall, Upper Saddle River, NJ, pp. 711–713.

Deichman, C.L., 1980. Genotype by soil fertility interactions at .01%BLSD. In: Proceedings, National Fertilizer Solutions Assoc. 1980 Summer Round Up, Chicago, IL.

Deichman, C.L., 2000. Plant arrangement for improving crop yields. U.S. Patent 6,052,941. Date issued April 25, 2000.

Deichman, C.L., 2002. Seminal corn yield results of the solar corridor corn production system III. In: Proceedings of ASA-CSSA-SSSA Annual Meetings, Indianapolis, IN.

Deichman, C.L., 2009. A sustainable farming system to maximize photosynthesis. In: Proc. Farm. Syst. Design Symp., Monterey, CA, pp. 87–88.

Deichman, C.L., Barrett, W., 1978. Heritability of corn hybrid response to plant nutrients. In: Proceedings of ASA-CSSA-SSSA Annual Meetings, Chicago, IL.

Deichman, C.L., Murrell, L., 1977. Differential corn hybrid responses to soil fertility variables. In: Proceedings of ASA-CSSA-SSSA Annual Meetings, Los Angeles, CA.

Deichman, C.L., Nelson, R.A., 2000. Novel solar corridor corn yield results by four Commercial Hybrids I. In: Proceedings of ASA-CSSA-SSSA Annual Meetings, Minneapolis, MN.

Deichman, C.L., Nelson, R.A., 2007. The solar corridor hypothesis. In: American Society of Agronomy Annual Meeting, New Orleans, LA, November 4–8, 2007.

Deichman, C.L., Warsaw, H., 1988. A novel research protocol to examine the source/sink relationship in corn. In: Proceedings of ASA-CSSA-SSSA Annual Meetings, Anaheim, CA.

Deichman, C.L., Kremer, R.J., Lupo, A., 2016. Over 1000 gallons of liquid biofuels per acre from four different corn hybrids on a Guilford fine sandy loam. In: Proceedings, International Annual Meetings, ASA-CSSA-SSSA, Phoenix, AZ.

Deichman, C.L., Nelson, R.A., Glanzer, D., 2001. Solar corridor corn yield results II. In: Agronomy Abstracts, ASA-CSSA-SSSA Annual Meetings, Charlotte, NC.

Kremer, R.J., Deichman, C.L., 2014. Soil quality and the solar corridor crop system. Agron. J. 106, 1853–1858.

Kremer, R.J., Deichman, C.L., Reinbott, T.M., Veum, K.S., 2015. Soil biology is enhanced through synergistic relationships in the solar corridor production system. In: Agronomy Abstracts. ASA-CSSA-SSSA Annual Meetings, Minneapolis, MN.

Nelson, K.A., 2014. Corn yield response to the solar corridor in upstate Missouri. Agron. J. 106, 1847–1852.

Nielsen, R.L., 2017. Historical corn grain yields for the U.S. <http://www.kingcorn.org/news/timeless/YieldTrends.html>.

Paterson, E., 2003. Importance of rhizodeposition in the coupling of plant and microbial productivity. Eur. J. Soil Sci. 54, 741–750.

Reinbott, T.M., Kremer, R.J., Deichman, C.L., 2018. Screening corn hybrids for the solar corridor production system. In: Agronomy Abstracts, ASA-CSSA Annual Meetings, Baltimore, MD.

Schmidt, J.E., Bowles, T.M., Gaudin, A.C.M., 2016. Using ancient traits to convert soil health into crop yield: impact of selection on maize root and rhizosphere function. Front. Plant Sci. 7, 373. <https://doi.org/10.3389/fpls.2016.00373>.

Smith, D., 2000. The solar corridor: patented 60" rows provide more sunlight, spike yields 40%. Farm J. 124 (9), special feature page 1.

Timlin, D., Kim, S.-H., Fleisher, D., Wang, Z., Reddy, V.R., 2019. Maize water use and yield in the solar corridor system: a simulation study. In: Deichman, C.L., Kremer, R. J. (Eds.), The Solar Corridor Crop System. Elsevier Academic Press, San Diego, CA.

CHAPTER 3

Maize water use and yield in the solar corridor system: a simulation study

Dennis Timlin[1], Soo-Hyung Kim[2], David Fleisher[1], Zhuangji Wang[3] and Vangimalla R. Reddy[1]
[1]USDA-ARS Adaptive Cropping Systems Laboratory, Beltsville, MD, United States
[2]University of Washington, Seattle, WA, United States
[3]University of Maryland, Plant Sciences and Landscape Architecture, College Park, MD, United States

Contents

Introduction

The Solar Corridor system is an agronomic approach to increase crop yields by using wide row spacings that allow light to penetrate the lower areas of the canopy thus increasing photosynthesis per unit ground area (Kremer and Deichman, 2014). Current agricultural practices, however, recommend narrow row spacing (Nelson, 2014) as this leads to earlier canopy closure that helps reduce weeds and increases early light interception. Wider row spacings, on the other hand, may have several advantages that include faster warming of soil in the spring as more soil is exposed to solar radiation early in the season (Kremer and Deichman, 2014). It is also suggested that wide rows lead to higher photosynthetic rates due to side

The Solar Corridor Crop System 2019 Published by Elsevier Inc.
DOI: https://doi.org/10.1016/B978-0-12-814792-4.00003-6

lighting of the canopy. Wide rows may also provide an increase in water availability when rainfall is limited (Lyon et al., 2009) due to a decrease in competition for water. Intercropping of a cover crop or other agronomic crop such as potato can also be done in wide rows providing the potential to increase net productivity per area (Hamdani and Suradinata, 2015).

Benefits of wide row spacing are not always consistent and often depend on growing conditions and variety. In research specifically targeted to the Solar Corridor concept, Nelson (2014) showed no difference in yield between 76 and 152 cm row spacings in 1 year and wider row spacing having lower yields in a dry year. In this same study, twin row corn with 152 cm spacing yielded higher at the high plant population. Maddonni and Martinez-Bercovich (2014) reported that 0.52 m row spacings had higher yields than 0.38 m using plant populations that varied from 45,000 to 68,000 plants ha^{-1}. The response to row spacing varied with landscape position due to differences in available water. See Chapter 2 for specific conditions leading to the results presented by Nelson (2014). On the other hand, corn yield increased by 4% in 2 out of 3 years when going from 0.76 m spacing to 0.51 row spacing in an irrigated study in Nebraska (Shapiro and Wortmann, 2006) with a plant population of 60–70,000 plants ha^{-1}. While row spacing effects are not consistent, increasing plant population from 35,000 to 120,000 plants h^{-1} generally increases yield (Nelson, 2014; Stone et al., 2000), but the response varies. The response of corn yield to plant population at a single site can range from parabolic to no response to a linear decline with increasing population in a dry year (Teasdale, 1998). Plant populations above about 120,000 generally result in decreasing yields (Stone et al., 2000). Overall, the response of grain yield to row spacing largely depends on water availability.

The effects of plant population, row spacing, and crop yields have been heavily researched over the years (Bruns and Abbas, 2003; Stone et al., 2000). Improved varieties and technology create new opportunities to revisit different plant arrangements as well as study interacting factors. Light capture between silking and grainfill is critical for yield formation, and plant arrangements that maximize this are most optimal (Westgate et al., 1997). As long as the plants are uniformly distributed, row spacing has little effect for the same population density (Maddonni et al., 2001,a,b). As plant population increases, kernel and ear weight, and leaf area per plant decrease and LAI increases (Bruns and Abbas, 2003). Narrowing of the leaves contributes to the smaller leaf area (Maddonni et al., 2001,a,b).

As a result of increasing LAI, in high population (or narrow spacing) the time to 95% light interception decreases. This increased light interception may not always result in increased yield however. Ear formation is delayed as population density increases (Jacobs and Pearson, 1991). High maize plant populations result in smaller seed. These have lower sink strength and cannot take advantage of the assimilate source from photosynthesis (Borrás et al., 2004). As a result the plants cannot fully utilize the increased intercepted light afforded by the more narrow row spacing (Maddonni et al., 2006). Even though plant size decreases, total biomass and grain yield per unit area increases with population density, but the rate of increase diminishes as population increases. Increases in grain yield with population density are more closely associated with increases in biomass per unit area than increases in harvest index (Sarlangue et al., 2007).

The response to plant density can also be variety specific. Some maize varieties can adapt to increasing plant density by changing the spatial arrangement of their leaves. Varieties that exhibit "plasticity" can modify leaf orientation to fill empty spaces due to plant arrangement. Plastic varieties in a rectangular configuration can increase or maintain light interception to maintain similar interception as in square arrangements (Maddonni et al., 2001,a,b). Thus a "rigid" variety would have less light interception in a rectangular configuration than in a more uniform, square configuration. The relative response of yield to plant density (i.e., an increase in yield as population density increases) is greater for short season varieties than for long season varieties (Sangoi, 2001; Sarlangue et al., 2007). Short season varieties produce low biomass and intercept less radiation at low population densities than longer season varieties (Sarlangue et al., 2007). This is due to low biomass plasticity of short season varieties, that is, they cannot increase ear size to change the harvest index as population density decreases. Harvest index decreases with plant population increases in short season varieties. At high populations, the reproductive sink capacity of the kernels in short season varieties is not limited as it is in low populations (because of smaller ears). Because they have larger ears, the late season varieties are not as sink limited in low plant densities as are short season varieties (Sarlangue et al., 2007). As a result, the rate of yield increase with increasing population density is less for late season varieties. Overall, long season varieties have higher yield potentials under low population than short season, and short season varieties have higher yield potential under high populations.

Water use efficiency (WUE) is also affected by plant population and row spacing. As population density increases, canopy closure occurs earlier thus reducing the contribution of soil evaporation to evapotranspiration (ET). Increasing plant population density (PPD) can increase WUE when water is not limiting as under irrigation (Ogola et al., 2005) through an increase in transpiration efficiency. As PPD increases, however, a larger canopy can increase competition for water and thus reduce WUE when water is limiting (Ogola et al., 2005). WUE is higher in narrow rows versus wider rows because water use increases more with increasing row space than yield does (Yao and Shaw, 1964). Barbieri et al. (2012) reported that reduced row spacing increased WUE under drought. Using drip irrigation in a sandy soil in Egypt, El-Hendawy et al. (2008) reported lower yields with increasing plant population under low levels of irrigation in sandy soils. The response was dependent on irrigation rates; yields in the medium density treatment were increased with higher irrigation rates. The optimal population was 48,000 plants ha^{-1}. The number of grains per ear and ear weight both decreased with increasing population but the number of grains per row of ear were consistent (El-Hendawy et al., 2008).

Plant population and row spacing also affect N use. Grain N was higher in narrow spacing (0.51 m) versus wide (0.76) in 2 out of 3 years in an irrigated study in Nebraska, United States (Shapiro and Wortmann, 2006). But N uptake was not affected by plant population. Nitrogen use efficiency (NUE) increases with decreasing row spacing because of better N use by the crop (Barbieri et al., 2008). When holding plant population constant, yield and WUE were higher on narrow row spacing (0.38 m) versus wide (0.76 m) due to increased light interception and more uniform leaf and root distribution (Sharratt and McWilliams, 2005). The relative response to narrow rows decreases and N availability decreases (Barbieri et al., 2008).

The effects of plant population and row spacing on N, WUE, and yield is complex involving many related processes. Climate and management (nutrients, tillage, planting date, and variety) affect the responses. Most field studies can only investigate a limited number of years of data and treatments. Crop simulation models are becoming important tools to investigate crop response to management and climate over long periods of time (Ahuja et al., 2006). Saseendran et al. (2005) used the Root Zone Water Quality Model and CERES-Maize to evaluate planting date effects on yields in Colorado for differing plant maturity groups. They determined the latest planting dates for the short, mid, and long season varieties to have a 50% chance of achieving a break-even yield under irrigation.

The CERES-Maize model was used to do an economic analysis of irrigation on maize yields (DeJonge et al., 2007). They showed that scheduled irrigation improved yields in 14 of the years simulated but economic returns were low. The objective of this study is to use the maize simulation model MAIZSIM (Kim et al., 2012) to investigate plant density and row spacing effects on maize yield, WUE, and nitrogen use for two contrasting soils and 30 years of weather data. MAIZSIM has a 2D finite element soil model that makes it well-suited for this purpose.

Materials and methods

Model

We used the maize model MAIZSIM that was developed by the USDA-ARS Adaptive Crops and Systems Lab at the Beltsville Agricultural Research Center in Beltsville, Maryland (Kim et al., 2012). The crop component of MAIZSIM simulates maize growth and development as a function of light, temperature, humidity, CO_2, nitrogen, water status, and soil properties (Kim et al., 2012). The maize model is interfaced with 2DSOIL, a soil process submodel that has a two-dimensional finite element representation of the soil profile and simulates heat and water movement, and solute transport. MAIZSIM also estimates ET and heat flux at the soil surface (Timlin et al., 2002) as well as root growth as taken from the soybean model GLYCIM (Acock et al., 1985). Temperature effects on growth are modeled using hourly ambient temperatures where rates are obtained from a peaked-exponential function (Yin et al., 2003) rather than growing degree days (GDD). Photosynthesis is simulated using a biochemical model of C_4 photosynthesis (Von Caemmerer, 2000) coupled with a stomatal conductance (Ball et al., 1987) and an energy balance model. Leaf-level photosynthesis is scaled to the plant level by applying the sunlit and shaded leaf approach (Campbell and Norman, 2012; De Pury and Farquhar, 1997; Lizaso et al., 2005).

The MAIZSIM model requires as input, daily or higher frequency values of total solar radiant energy (MJ), ambient temperature (°C), precipitation (mm), relative humidity (%), and latitude and longitude (to calculate day length and sun angle). If only daily weather data are available, they can be scaled to hourly values internally in the model (Timlin et al., 2002). Required soils data include soil texture, bulk density (g cm^{-3}), saturated hydraulic conductivity (cm day^{-1}), and water retention curve parameters (Van Genuchten, 1980). The current version requires two variety specific parameters: the generic minimal number of leaves at maturity (Ng) and a Stay-Green parameter. Stay-Green is a variety characteristic that is related to how rapidly the plant leaves senescence. At higher values of Stay-Green the plant leaves stay greener for a longer time. The juvenile leaf number at maturity is normally available in standard variety data from seed companies. The values for these parameters will be discussed later in this section. We used, as *k*-factors, 0.40 and 0.45 for the 1.25 and 0.75 m spacings in order to calculate light interception using LAI and Beer's law (Flenet et al., 1996) for partitioning ET to E and T.

The effects of plant population on leaf growth were modeled as a function of the red to far-red light ratio in the canopy (Maddonni et al., 2001,a,b; Zhu et al., 2014). Although leaf initiation rates were shown to be affected by plant population (Zhu et al., 2014), the difference was not large, and so this was not incorporated into the model. There is evidence that the red:far-red ratio of light is affected by the relative shading by neighboring plants (Maddonni et al., 2001,a,b; Zhu et al., 2014). This shade effect was then modeled as a function of the ratio of the areas of sunlit to shaded leaves where the relative areas are determined by equations from Campbell and Norman (2012). We used the daily mean ratio of shaded to sunlit leaves. A three-point sigmoidal function (Eq. 3.1) was chosen to represent the effect (Fig. 3.1).

$$E_s = \min 1, \left(\frac{A}{1 + e^{-((\mathrm{SR} - X_o)/B)}} \right) \tag{3.1}$$

Where SR is the sunlit ratio (sunlit area/[sunlit area + shaded area]), $A = 2.0$, $X_o = 0.85$, and $B = 0.65$. The result (a number from 0 to 1) is multiplied by the leaf expansion rate to reduce leaf growth if the number is less than 1. At common plant densities, the value is about 0.86. The parameters were fit using the data in Zhu et al. (2014). The implementation was tested against data from a population study in the southeastern United States (Bruns and Abbas, 2003). The yields were obtained from their Fig. 1. They used five plant populations ranging from 43,000 to 76,500 plants ha^{-1}. Weather and soil data for Mississippi were not available so we used data from the current study as described later.

Simulations

We chose Iowa as a location because of the importance of corn in that state. Soils and weather data were from Johnson County, Iowa. Two soils were used: the first was a Wiota silty clay loam, which is classed as a

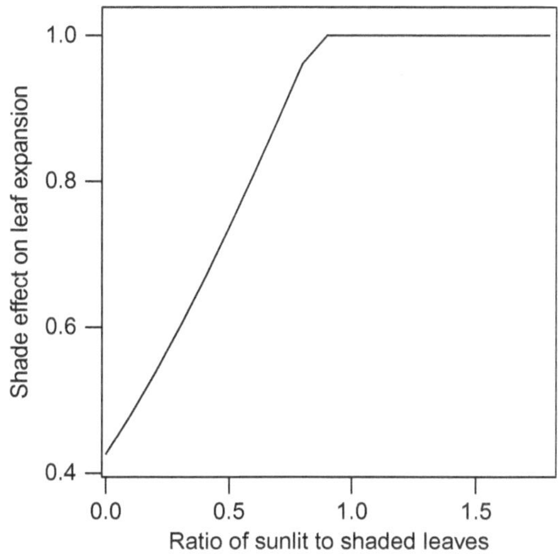

Figure 3.1 Effect of leaf shading due to population density on leaf growth. The data are from Eq. (3.1). The shade effect is used to adjust leaf growth.

fine-silty, mixed, superactive, mesic Pachic Argiudoll and represents a soil with high water-holding capacity. The other soil was a Chelsea loamy fine sand, which is classified as a mixed, mesic Lamellic Udipsamment and represents a well-drained soil with low water-holding capacity. The soil textural properties (Table 3.1) were obtained from the NRCS Soils SSURGO database (Soil Survey Staff, 2018). The program Rosetta (Schaap et al., 2001) was used to generate saturated soil hydraulic conductivity and parameters for the van Genuchten equation (Van Genuchten, 1980) using these textural values and bulk density (Table 3.2). The weather data came from a simulated 30-year sequence of daily radiation (MJ), maximum and minimum temperatures (°C), precipitation (mm), and average wind speed (km $hour^{-1}$). The weather simulator CliGen V 4.2 (Nicks and Gander, 1994; Nicks et al., 1995) was used to simulate the weather using parameters from Iowa City, Iowa. Planting dates were calculated as the day when 100 GDD (base temperature of 8°C accumulated from February 1 of that year). Most planting dates were in mid-April. The management details are listed in Table 3.3. The finite element grid used a seepage face at the bottom of the domain. The bottom of the domain was set at 2.00 m to minimize the impact of soil depth on water availability (profile depth was 2 m). The effect of this boundary condition is that water will build up at the bottom boundary until it reaches saturation and then drains at the rate of the saturated hydraulic conductivity multiplied by the gradient.

Table 3.1 Soil physical properties and initial conditions used in the simulations.

Soil	Bottom depth of layer (m)	OM (%)	Initial N (mg L^{-1})	Initial soil matric potential (cm)	Initial soil temperature (°C)	Sand (%)	Silt (%)	Clay (%)	Bulk density (g cm^{-3})
Chelsea	0.20	1.5	30	−100	23	75	15	10	1.50
	0.91	0.5	13	−100	23	80	15	5	1.60
	2.00	0.5	2	−100	23	80	15	5	1.60
Wiota	0.20	3.0	40	−100	23	15	60	25	1.34
	0.56	2.0	22	−100	23	17	51	32	1.34
	1.22	1.0	5	−100	23	20	60	20	1.50
	2.00	0.5	5	−100	23	65	25	10	1.60

Table 3.2 Soil hydraulic properties for the two soils used in the simulations.

Soil	Depth (m)	Residual water content (cm^3 cm^{-3})	Saturated water content (cm^3 cm^{-3})	Alfa (cm^{-1})	*n*	Saturated hydraulic conductivity (cm d^{-1})
Chelsea	0.20	0.051	0.396	0.082	1.85	120
	0.91	0.051	0.43	0.082	1.75	87
	2.00	0.031	0.38	0.069	1.8	60
Wiota	0.20	0.010	0.38	0.03	1.8	65
	0.56	0.01	0.28	0.03	1.8	23
	1.22	0.01	0.28	0.03	1.8	23
	2.00	0.051	0.396	0.082	1.85	120

Table 3.3 Management practices used in the simulations. Nitrogen was split into two applications. The first was applied 5 days before sowing and the second applied 30 days after sowing.

Row spacing (m)	Population (plants ha^{-1})	Nitrogen (kg ha^{-1} N)	Planting date
1.00	37,000	225	Mid-April to early May
0.76	68,000	225	Mid-April to early May

Table 3.4 Parameter values for the maize variety parameters used in the simulations.

Variety	Juvenile leaves (Ng)	Stay-Green
Short	13	2
Medium	16	4.5
Long	20	6

We considered three types of corn varieties represented by contrasting relative maturity (RM) values. RM was simulated by using crop parameters that controlled Stay-Green properties and juvenile leaf number. Juvenile leaf number is the default number of leaves a corn plant has before it tassels. Both of these values can be considered to have a variety of genetic properties. Final leaf number can increase depending on temperature and day length (Grant, 1989). Since the number of leaves determines when a corn plant will tassel, it can be used as an indicator of RM. The parameters used for MAIZSIM are listed in Table 3.4.

Validation for plant population effects

We validated MAIZSIM by carrying out indirect tests of the model to determine if the simulated yields were realistic for similar conditions. We investigated the range of simulated yields against those from Johnson County, Iowa (www.nass.usda.gov/Statistics_by_State/Iowa/index.php).

Statistical analysis

The simulation results were analyzed using PROC Mixed of SAS (Version 9.4 of the SAS system for Windows © 2017, SAS Institute Inc.). The analysis used population, spacing, and variety as treatments. Year × soil was treated as a repeated variable. The model has been recently developed, and the code and input files for this chapter are available in GitHub (https://github.com/ARS-CSGCL-DT/MAIZSIM).

Results

The first test was to evaluate the model's response to plant population using yield data from Mississippi (Bruns and Abbas, 2003). MAIZSIM was able to realistically simulate the relative effects of plant population on grain yields (Fig. 3.2). The relative magnitudes of the grain yields are different from the published values because of potential differences in location, weather, and variety. The slopes of the relationship between population and yield are very similar (5.556×10^{-5} kg ha^{-1} plant^{-1} for simulated vs 5.0×10^{-5} for the Mississippi data). The average yields of the model for the 30-year period were compared to the average yield in Johnson County, Iowa over the past 20 years (Fig. 3.3). The simulated average simulated grain yield for all years is 12.02 Mt ha^{-1}, which is close to the county average of 10.04 for Johnson County, Iowa, from which the weather and soil data were taken. The ranges are within the standard deviations of both groups (Fig. 3.3). The simulated yields are expected to be slightly higher since the simulations represent optimum conditions.

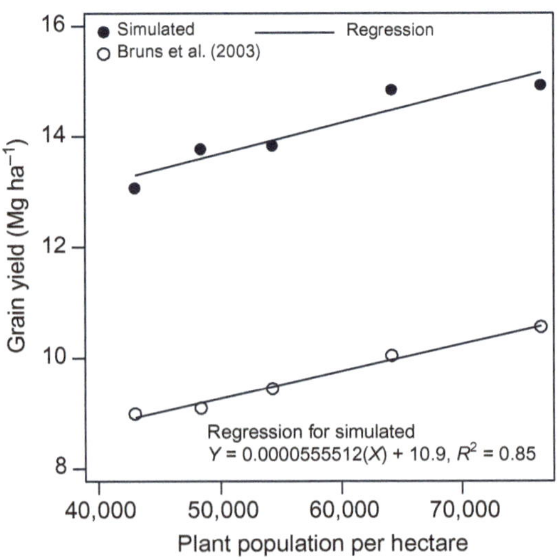

Figure 3.2 Simulated and observed yield response to plant population. Observed data are from Bruns and Abbas (2003). Simulations were run with 4 years of weather data from the Iowa site using the same row spacing and plant populations as in Bruns and Abbas (2003).

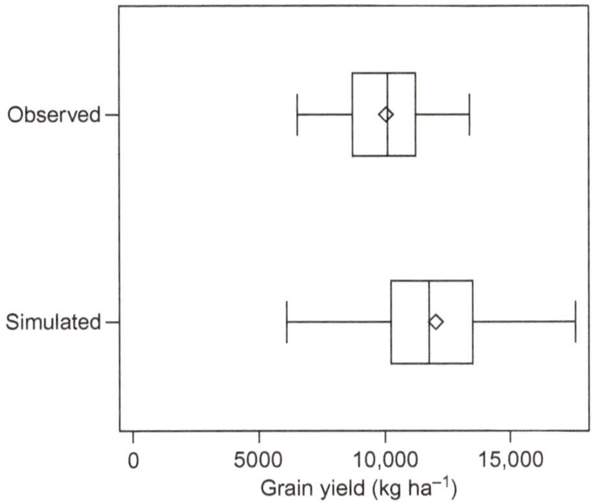

Figure 3.3 Mean yields and their distributions for yields from Johnson County, Iowa, and simulated yields. The observed yields were from the years 1995 to 2017. Thirty years of simulated yields were used.

Table 3.5 Mean time to anthesis and time from anthesis to maturity for the three varieties.

Variety	Time to anthesis		Time from anthesis to maturity	
	Days	GDD (°C)[a]	Days	GDD (°C)
Short	47	500	40	536
Medium	58	641	86	927
Long	72	837	97	772

[a]Base temperature of 10°C.

The simulated average times to anthesis and maturity are given in Table 3.5. Times to anthesis vary from 47 to 72 days with the longer season varieties having longer periods of juvenile growth. The total growing period varies from 87 to 169 days for the different maturity groups. The number of accumulated GDD generally increased with days to anthesis and maturity as the variety went from short season to long season. The only exception was the long season variety which accumulated less GDD than the medium season even though it took a longer time in days to reach maturity from anthesis. This is because the late season variety reached anthesis later in the season than the medium season variety did. Not as many GDD were accumulated due to lower temperatures.

The tests for significance for the main treatment effects are given in Table 3.6. For most of the variables, only variety and population had significant effects. There was a significant effect of spacing and its interaction with variety only on seasonal photosynthetically active radiation (PAR) interception. Population, variety, and their interaction had a significant effect on all the variables except potential soil evaporation and the transpiration ratio. Population did not have a significant overall effect on WUE, but the significant interaction with variety means there was an effect for one or more of the varieties.

The medium RM variety had the highest yields (Table 3.7) while the low RM (short season) variety had the lowest. The yields for the high RM variety were slightly less than the yield for the midseason though biomass was significantly higher. This is likely due to the lower temperatures later in the season. LAI and leaf area increased with RM, which is expected due to the higher number of leaves in the longer season varieties. The potential and actual transpiration values also increased with RM and growing season length. Evaporation of water from the soil decreased with RM probably because of the increased crop cover with increasing RM (Table 3.7). The values for transpiration ratio were very similar though the differences were significant. The values close to one indicate little overall water stress. The average transpiration ratio decreased with RM likely due to the longer growing season with more growth in the later, dryer time periods. WUE also decreased with increasing RM.

The details of the means in Table 3.8 show there were no significant differences between the two row spacings for most of the variables. The only exception was slightly higher intercepted PAR for the 0.75 m row spacing than for the 1.20 m spacing for the mid and early season varieties. This did not result in a significant difference in yield or dry matter. Except for the long season variety, grain yields were higher in the high population treatments than in the low one. LAI was higher in the high population treatments, but leaf area per plant was lower. All ET variables except for soil evaporation were higher in the high population treatments than in the low ones. The effects on WUE varied with population and variety. The WUE was higher in the high population treatments for the low RM variety, was not affected by population in the medium RM variety, and was less in the high RM variety for the high population treatment. The differences in crop cover were larger for the early RM variety than the others which likely contributed to the higher WUE for high population. Intercepted PAR was consistently larger for the high

Table 3.6 Tests of significance for treatment effects.

Source of variation	*F*-Value													
	Grain	WUE	Cover	Dry matter	LAI	Transpiration ratio	Potential trans	Actual trans	Potential soil Ev	Actual soil Ev	PET	AET	Leaf area	Intercepted PAR
Variety	198.5***[a]	787.8***	34980***	1088***	90397***	66.02***	10507***	1324***	15.77***	378.3***	8235***	1113***	95668***	119E3***
Pop	68.76***	1.24	41670***	259.6***	116E3***	31.84***	2548***	273.1***	29.65***	634.2***	1465***	180.6***	1589***	11560***
Pop × variety	9.97***	39.54***	2528***	5.02*	4495***	0.03	103.9***	7.47*	1.76	20.19***	128.0***	8.25*	94.10***	341.5***
Pop × space	0.07	0.02	0.01	0.25	0.02	0.36	0.15	0.34	0.00	0.00	0.19	0.32	0.03	0.87
Pop × space × variety	0.05	0.01	0.00	0.05	0.03	0.11	0.08	0.09	0.00	0.01	0.08	0.08	0.01	0.21
Space	0.28	0.03	0.12	0.79	0.01	2.43	0.80	1.58	0.63	0.12	0.93	0.79	0.03	337.6***
Space × variety	0.01	0.29	0.18	0.08	0.19	0.12	0.14	0.24	0.00	0.00	0.22	0.22	0.18	106.2***

[a]Indicates $P < 0.05$; ***Indicates $P < 0.001$.

Table 3.7 Main effects on agronomic variables averaged over variety (relative maturity).

Variety	Grain (kg ha^{-1})	WUE (kg m^3)	Cover	Dry matter (kg ha^{-1})	LAI	Transpiration ratio	Potential trans (mm)	Actual trans (mm)	Potential soil, *E* (mm)	Actual soil, *E* (mm)	PET (mm)	AET (mm)	Actual ET (mm)	Leaf area (cm^2 plant^{-1})	Intercepted PAR (mol m^{-2})
Short (low RM)	8.3c*	3.0a	0.79c	11.8c	2.0c	0.96a	196.7c	187.9c	185.3b	86.9b	383.4c	275.1c	275.1c	4142c	295.2c
Mid (medium RM)	12.1a	2.6b	0.89b	19.5b	3.0b	0.89b	424.6b	378.5b	186.6a	91.6a	617.7b	471.7b	471.7b	6077b	661.1b
Long (high RM)	11.0b	1.9c	0.95a	22.4a	4.2a	0.87c	558.8a	486.8a	159.9b	78.4b	729.4a	566.2a	566.2a	8481a	1060a

*Within columns and varieties, means followed by the same letter are not significantly different according to LSD ($P < 0.01$).

Table 3.8 Effects of population and row spacing on agronomic outputs averaged over varieties and aggregated by relative maturity (RM) for 30 years of weather conditions.

Variety	Population	Row spacing (m)	Grain (kg ha^{-1})	WUE (kg m^3)	Cover	Dry matter (kg ha^{-1})	LAI	Transpiration ratio	Potential transp (mm)	Actual transp (mm)	Potential soil, *E* (mm)	Actual soil, *E* (mm)	PET (mm)	AET (cm)	Leaf area (cm^2 plant^{-1})	Intercepted PAR (mol m^{-2})
Short	Low	0.75	7.3b*	2.9b	0.71b	10.2b	1.6b	0.97ab	165.2b	160.8b	198.4a	94.3a	364.3b	255.3b	4245a	256.9c
	High	0.75	9.2a	3.1a	0.86a	13.3a	2.5a	0.93b	227.9a	213.4a	172.5b	81.1bc	402.5a	295.0a	4047b	358.4a
	Low	1.20	7.3b	2.9b	0.71b	10.3b	1.6b	0.98a	165.3b	161.7b	198.1a	92.7ab	364.2b	254.6b	4236a	234.3d
	High	1.20	9.3a	3.1a	0.86a	13.4a	2.5a	0.94ab	228.4a	215.5a	172.2b	79.6c	402.7a	295.5a	4039b	331.0b
Medium	Low	0.75	11.2b	2.6a	0.84b	17.6b	2.3b	0.90a	365.7b	332.4b	200.3a	99.1a	570.1b	432.8b	6229a	593.2c
	High	0.75	12.8a	2.5a	0.94a	21.2a	3.7a	0.86b	481.2a	417.3a	173.3b	85.8bc	662.5a	504.7a	5922b	774.6a
	Low	1.20	11.2b	2.6a	0.84b	17.6b	2.3b	0.91a	366.2b	334.1b	200.0a	97.3ab	570.6b	432.9b	6229a	548.3d
	High	1.20	13.1a	2.5a	0.94a	21.6a	3.7a	0.88ab	485.5a	430.3a	172.8b	84.2c	667.4a	516.3a	5927b	728.3b
Long	Low	0.75	10.8a	2.1a	0.92b	21.2b	3.2b	0.88ab	491.5b	436.1b	167.3a	81.9a	665.3b	518.7b	8715a	993.1b
	High	0.75	11.1a	1.8b	0.98a	23.4a	5.1a	0.84b	623.3a	528.2a	152.7b	76.4a	789.8a	605.6a	8245b	1127a
	Low	1.20	10.8a	2.0a	0.92b	21.3b	3.2b	0.89a	494.1b	443.2b	166.9a	80.4a	668.4b	524.4b	8715a	993.7b
	High	1.20	11.2a	1.8b	0.98a	23.7a	5.1a	0.85ab	626.4a	539.7a	152.6b	75.0a	794.1a	615.9a	8249b	1126a

*Within columns and varieties, means followed by the same letter are not significantly different according to LSD ($P < 0.01$).

population treatments than for the low. The high population treatments were also more likely to have water stress as indicated by the lower transpiration ratios. Crop cover was also higher in the high population treatment leading to higher rates of transpiration.

Discussion

There was no effect of row spacing on most of the agronomic variables. This is consistent with field research results (Maddonni et al., 2001,a,b; Stone et al., 2000). When population is kept constant, row spacing does not generally affect LAI (Liu et al., 2012) except at very low populations (Maddonni et al., 2001,a,b). However, Nelson (2014) did report that yield was higher in 0.76 m versus 1.52 m row spacing, but in only 1 year. Maddonni et al. (2001,a,b) reported row spacing differences between 0.35 and 0.70 cm, but only at very low populations ~ 3 plants m^{-2}.

The increase in individual plant leaf area and decrease in LAI with increasing row spacing is a common effect of decreasing population density (Subedi et al., 2006). Plants grow larger to capture the available light, but LAI is less because LAI is a plant per unit area dependent value. LAI is small in the low population treatments and this results in a high ratio of sunlit to shaded leaves. The higher leaf area per plant can support high canopy photosynthetic rates in spite of low LAI. The high sunlit ratio, however, provides for more transfer of light through the canopy and there is less diffuse light and scattering within the canopy. As a result, the light is not used efficiently. On the other hand, when the plant density is high, although the total leaf area per plant is smaller, there is more complete interception of light between rows and better use of scattered and diffuse light. This contributes to the higher light interception and yields. The differences in light interception between low and high density plantings in general are not constant throughout the day (Timlin et al., 2014). The differences are greatest at midday in the midrow when the sun is directly overhead and less in the morning and evening when the sun angle is lower. The differences are also less in the plant row (stem location). Increased plant population also increases leaf senescence rate (Borrás et al., 2003), most likely due to reduced assimilate availability for grain and increased water and nitrogen stress. Borrás et al. (2003) did not find an effect of light quality (red:far-red ratio) on senescence. MAIZSIM does account for some of these processes in senescence (nutrient and water stress) but not the effect of plant population.

The simulation results showed a significant positive response of yield to plant population for the early- and medium-RM varieties. Stone et al. (2000) looked at the effects of plant population on two varieties in New Zealand with varying RM. Both varieties responded to plant population, but, in contrast to our study the rate of increase of yield was higher for the later season variety. The populations in this study, however, were about half that in the Stone et al. (2000) study. In a study in Minnesota, a 101-day variety had higher yields than a 95-day variety but a third variety with a 105-day RM value was not different than a 101-day variety (Van Roekel and Coulter, 2012). Assefa et al. (2016) reported similar results and showed the relationships with latitude. At latitudes above 40 degree where Iowa is located, there is only a small change in yield as population increases above 60,000 plants ha^{-1} for longer season varieties. There is also little difference in the response to population between the very long season varieties (CRM > 115) and medium season varieties (CRM 106−105) at this latitude. In our study, the responses were similar, about 2 Mt ha^{-1} between the two plant populations (Figure 4 in Assefa et al., 2016). Maximum yield at higher plant densities overall depends on the yield environment (i.e., temperature and solar radiation). In high yielding environments, yields can continue to increase for plant populations approaching 100,000 plants ha^{-1} (Assefa et al., 2016). Overall, the conclusions of the study by Van Roekel and Coulter (2012) were the same as in this study—high yields could be obtained from mid and late-RM varieties with high plant density.

The use of an exponential layering model that uses a homogenous, layered LAI model for canopy light interception may not be optimal to account for solar movement during the day and wide spacing between rows. Models with more comprehensive 3D architecture incorporating components such as heterogenous canopies, position relative to the sun angle and leaf angle may provide a more accurate representation of light interception in row crops (Sarlikioti et al., 2011). The light interception model in MAIZSIM does account for leaf shape but not leaf angle (Campbell and Norman, 2012) nor the angle of the row relative to the solar path. MAIZSIM does not account for changes in light interception as the solar angle changes for the purposes of photosynthesis calculations but not for calculation of soil cover. We did use different light transmission factors for the two population densities (Flenet et al., 1996) to calculate soil cover, but they did not change during the day.

There are many plant qualities and behaviors that contribute to improved performance of maize with increased (or decreased) plant population, but most current models do not account for them. For example, maize varieties can perform better under high plant population if there is a more compact canopy structure and lower ear placement (Subedi et al., 2006), a characteristic of early maturing varieties. Flex-ear maize varieties can adjust kernel number based on growing conditions. These varieties tend to produce an ear with more kernels at low plant density than do varieties with a fixed kernel number (Abbas et al., 2012). Maize plants are also able to adjust to the light environment to vary their leaf angles based on their perceptions of neighboring plants (Maddonni et al., 2001,a,b). Choosing varieties that are better able to adjust may be an option to increase yields in the Solar Corridor. Improving models to address these would also make them more useful for agronomic assessments.

Water use efficiency

WUE can be increased in temperate and humid environments by increasing plant population (Hatfield et al., 2001). In our study WUE was higher for the high population than for the low one, but only for the low RM variety; spacing did not make a difference for any of the varieties (Tables 3.6 and 3.8). In all cases, cumulative soil evaporation decreased with increasing population due to increased crop cover and likely, more rapid closure of the canopy. For the medium and long RM varieties the canopy covered more than 85% of the soil for all population treatments and so soil evaporation would be a smaller component of total ET. Thus relative reductions in soil evaporation due to population differences would not contribute to increases in WUE for the longer season varieties. In a study in Texas, Howell et al. (1998) compared WUE for short and long season varieties. Based on their reported dates for planting and harvest, their varieties corresponded to our medium and long season varieties. Their WUE values for grain were about half our values, but the ET in that area was reported to be almost twice the values here. Even though there were differences in seasonal ET between the two varieties on the order of 120 mm, the WUE values were similar, about 1.65 kg m^{-3} because of increased yield in the long season variety. Both varieties had high LAIs so there was no advantage to a variety until grainfill where the short season variety senesced earlier. These are consistent with our results.

The response of WUE to population is not always consistent. Over a 3-year period using different irrigation treatments and plant populations from 57,000 to 81,000 plants ha^{-1}, Al-Kaisi and Yin (2003) reported only a few significant differences. Where the differences were significant, often the WUE increased with increasing population, especially between 57 and 69,000 plants ha^{-1}. Ogola et al. (2005) reported that high population decreased water loss via soil evaporation. WUE was higher for the high population treatment in their study, but only under irrigation. Water stress reduced WUE in the nonirrigated treatment. Interplant competition for water would increase the likelihood of water stress as plant population increases if there is not sufficient rainfall. In our study the transpiration ratio was highest for the short season variety, so it is likely that water stress may have contributed to the lack of difference in WUE between populations in the later season varieties.

In addition to what has been described previously there are other limitations to the modeling approach we have taken and how it applies to the Solar Corridor system. The light interception component (use of LAI and the sunlit/shaded grouping for photosynthesis) of the model is largely 1D and hence does not account well for row orientation and edge effects. Another factor is that MAIZSIM currently lacks the ability to correctly simulate source—sink relationships and partitioning. At this time, there is no code for kernel development in MAIZSIM, only partitioning of carbon to the ear is simulated. The amount of carbon sent to the ear is not limited by the size of the ear directly. Ear size is a function of the canopy photosynthetic rate, which is determined by plant size. Thus small plants will produce small ears and large plants large ears. It would be more realistic to begin calculating kernel number from the time the ear first initiates so that environmental conditions can influence the number and size of the kernels. The kernels then would become a realistic sink for photosynthate. There are some positive aspects of the model as well. The 2D soil model does differentiate soil temperature and soil water evaporation in the row and interrow zones. To some extent, the sunlit/shaded abstraction will account for the variance of direct beam radiance over the day and should provide reasonable estimates of photosynthesis when a wide row crop is radiated on the edges. Overall, the row spacing and plant population effects on yield and WUE simulated by MAIZSIM do represent results from field experiments.

Summary and conclusions

The Solar Corridor concept attempts to maximize the capture of solar radiation to increase yield for improved economic returns. Optimal capture of solar radiation is a complex process and depends on several factors. The goal of this study was to use a crop simulation model for maize to investigate the interactive effects of plant population (3.8 and 6.8 plants m^{-2}), row spacing (0.75 and 1.20 m) and RM of maize varieties on yield, WUE, and light interception. The simulations showed that row spacing did not significantly affect yield, WUE, or transpiration. For the same plant population, intercepted light was lower for the wider row spacing. These results were similar to those found in the literature. Yields were higher for the high population treatments but only for the short and midseason varieties. Yields for the long season variety were not different from the midseason variety suggesting that the longer period of grainfill for the long season variety was not an advantage because of decreasing day length and cooler temperatures. There were only a few years of weather data with strong water stress. When filtered for these years, there was a slight yield advantage to the wider spacing. This research suggests that, for the conditions used here, wider row spacing (to 1.2 m) and higher plant populations with a medium season variety generally result in higher yields. If water availability is limited, a short season variety would be preferable. Based on the results in this study, to maximize benefit from wide row spacing, a second crop in the midrow space would be beneficial. Many additional permutations of row spacing, varieties, soils, populations, and weather exist that could be investigated. Additional plant populations and interactions with irrigation could be investigated in the future. This chapter has shown that a simulation model can provide realistic simulations of maize growth in the Solar Corridor. The ability to simulate 3D plant architecture would greatly enhance the usefulness of the model as well as the ability to simulate an intercrop or twin crops.

References

Abbas, H.K., Mascagni Jr, H.J., Bruns, H.A., Shier, W.T., 2012. Effect of planting density, irrigation regimes, and maize varieties with varying ear size on yield, and aflatoxin and fumonisin contamination levels. Am. J. Plant Sci. 3, 1341.

Acock, B., Reddy, V., Whisler, F., Baker, D., McKinion, J., Hodges, H., et al., 1985. The soybean crop simulator GLYCIM: model doc, 1982. PB85 171163.

Ahuja, L.R., Ma, L., Timlin, D.J., 2006. Trans-disciplinary soil physics research critical to synthesis and modeling of agricultural systems. Soil Sci. Soc. Am. J. 70, 311–326. Available from: https://doi.org/10.2136/sssaj2005.0207.

Al-Kaisi, M.M., Yin, X., 2003. Effects of nitrogen rate, irrigation rate, and plant population on corn yield and water use efficiency. Agron. J. 95, 1475–1482. Available from: https://doi.org/10.2134/agronj2003.1475.

Assefa, Y., Vara Prasad, P.V., Carter, P., Hinds, M., Bhalla, G., Schon, R., et al., 2016. Yield responses to planting density for US modern corn hybrids: a synthesis-analysis. Crop Sci. 56, 2802–2817. Available from: https://doi.org/10.2135/cropsci2016.04.0215.

Ball, J.T., Woodrow, I.E., Berry, J.A., 1987. A model predicting stomatal conductance and its contribution to the control of photosynthesis under different environmental conditions. In: Biggins, J. (Ed.), Progress in Photosynthesis Research: Volume 4 Proceedings of the VIIth International Congress on Photosynthesis Providence, Rhode Island, USA, August 10–15, 1986. Springer, Netherlands, Dordrecht, pp. 221–224.

Barbieri, P.A., Echeverría, H.E., Saínz Rozas, H.R., Andrade, F.H., 2008. Nitrogen use efficiency in maize as affected by nitrogen availability and row spacing. Agron. J. 100, 1094–1100. Available from: https://doi.org/10.2134/agronj2006.0057.

Barbieri, P., Echarte, L., Della Maggiora, A., Sadras, V.O., Echeverria, H., Andrade, F.H., 2012. Maize evapotranspiration and water-use efficiency in response to row spacing. Agron. J. 104, 939–944. Available from: https://doi.org/10.2134/agronj2012.0014.

Borrás, L., Maddonni, G.A., Otegui, M.E., 2003. Leaf senescence in maize hybrids: plant population, row spacing and kernel set effects. Field Crops Res. 82, 13–26. Available from: https://doi.org/10.1016/S0378-4290(03)00002-9.

Borrás, L., Slafer, G.A., Otegui, M.A.E., 2004. Seed dry weight response to source–sink manipulations in wheat, maize and soybean: a quantitative reappraisal. Field Crops Res. 86, 131–146. Available from: https://doi.org/10.1016/j.fcr.2003.08.002.

Bruns, H., Abbas, H., 2003. Effects of plant populations on maize hybrids in the sub-tropical mid South USA [*Zea mays* L.]. Maydica 48, 21–27.

Campbell, G.S., Norman, J.M., 2012. An Introduction to Environmental Biophysics. Springer Science & Business Media.

De Pury, D.G.G., Farquhar, G.D., 1997. Simple scaling of photosynthesis from leaves to canopies without the errors of big-leaf models. Plant Cell Environ. 20, 537–557.

DeJonge, K.C., Kaleita, A.L., Thorp, K.R., 2007. Simulating the effects of spatially variable irrigation on corn yields, costs, and revenue in Iowa. Agric. Water Manage. 92, 99–109. Available from: https://doi.org/10.1016/j.agwat.2007.05.008.

El-Hendawy, S.E., El-Lattief, E.A.A., Ahmed, M.S., Schmidhalter, U., 2008. Irrigation rate and plant density effects on yield and water use efficiency of drip-irrigated corn. Agric. Water Manage. 95, 836–844. Available from: https://doi.org/10.1016/j.agwat.2008.02.008.

Flenet, F., Kiniry, J.R., Board, J.E., Westgate, M.E., Reicosky, D.C., 1996. Row spacing effects on light extinction coefficients of corn, sorghum, soybean, and sunflower. Agron. J. 88, 185–190.

Grant, R.F., 1989. Simulation of maize phenology. Agron. J. 81, 451–457. Available from: https://doi.org/10.2134/agronj1989.00021962008100030011x.

Hamdani, J.S., Suradinata, Y.R., 2015. Effects of row intercropping system of corn and potato and row spacing of corn on the growth and yields of Atlantic potato cultivar planted in medium altitude. Asian J. Agric. Res. 9, 104–112.

Hatfield, J.L., Sauer, T.J., Prueger, J.H., 2001. Managing soils to achieve greater water use efficiency. Agron. J. 93, 271–280. Available from: https://doi.org/10.2134/agronj2001.932271x.

Howell, T.A., Tolk, J.A., Schneider, A.D., Evett, S.R., 1998. Evapotranspiration, yield, and water use efficiency of corn hybrids differing in maturity. Agron. J. 90, 3–9. Available from: https://doi.org/10.2134/agronj1998.00021962009000010002x.

Jacobs, B.C., Pearson, C.J., 1991. Potential yield of maize, determined by rates of growth and development of ears. Field Crops Res. 27, 281–298. Available from: https://doi.org/10.1016/0378-4290(91)90067-6.

Kim, S.-H., Yang, Y., Timlin, D.J., Fleisher, D.H., Dathe, A., Reddy, V.R., et al., 2012. Modeling temperature responses of leaf growth, development, and biomass in maize with MAIZSIM. Agron. J. 104, 1523–1537. Available from: https://doi.org/10.2134/agronj2011.0321.

Kremer, R.J., Deichman, C.L., 2014. Introduction: The solar corridor concept. Agron. J. 106, 1817–1819.

Liu, T., Song, F., Liu, S., Zhu, X., 2012. Light interception and radiation use efficiency response to narrow-wide row planting patterns in maize. Aust. J. Crop Sci. 6, 7.

Lizaso, J., Batchelor, W., Boote, K., Westgate, M., 2005. Development of a leaf-level canopy assimilation model for CERES-Maize. Agron. J. 97, 722–733.

Lyon, D.J., Pavlista, A.D., Hergert, G.W., Klein, R.N., Shapiro, C.A., Knezevic, S., et al., 2009. Skip-row planting patterns stabilize corn grain yields in the central great plains. Crop. Manag. 8.

Maddonni, G.A., Martinez-Bercovich, J., 2014. Row spacing, landscape position, and maize grain yield. J. Int. J. Agron. 2014, 12. Available from: https://doi.org/10.1155/2014/195012.

Maddonni, G.A., Otegui, M.E., Cirilo, A.G., 2001a. Plant population density, row spacing and hybrid effects on maize canopy architecture and light attenuation. Field Crops Res. 71, 183–193. Available from: https://doi.org/10.1016/S0378-4290(01)00158-7.

Maddonni, G.A., Chelle, M., Drouet, J.-L., Andrieu, B., 2001b. Light interception of contrasting azimuth canopies under square and rectangular plant spatial distributions: simulations and crop measurements. Field Crops Res. 70, 1–13.

Maddonni, G.A., Cirilo, A.G., Otegui, M.E., 2006. Row width and maize grain yield. Agron. J. 98, 1532–1543. Available from: https://doi.org/10.2134/agronj2006.0038.

Nelson, K.A., 2014. Corn yield response to the solar corridor in upstate Missouri. Agron. J. 106, 1847–1852. Available from: https://doi.org/10.2134/agronj2012.0326.

Nicks, A.D., 1994. ter resouCLIGEN: a weather generator for climate inputs to water resource and other models. In: Proc. Fifth Int. Conf. on Computers in Agriculture.

Nicks, A., Lane, L., Gander, G., 1995. Weather generator. USDA-water erosion prediction project hillslope profile and watershed model documentation. NSERL Report: 2.1-2.22.

Ogola, J., Wheeler, T., Harris, P., Soil, 2005. Water use of maize in response to planting density and irrigation. S. Afr. J. Plant 22, 116–121.

Sangoi, L., 2001. Understanding plant density effects on maize growth and development: an important issue to maximize grain yield. Ciência Rural 31, 159–168.

Sarlangue, T., Andrade, F.H., Calviño, P.A., Purcell, L.C., 2007. Why do maize hybrids respond differently to variations in plant density? Aron. J. 99, 984–991. Available from: https://doi.org/10.2134/agronj2006.0205.

Sarlikioti, V., de Visser, P.H.B., Marcelis, L.F.M., 2011. Exploring the spatial distribution of light interception and photosynthesis of canopies by means of a functional–structural plant model. Annal. Bot. 107, 875–883. Available from: https://doi.org/10.1093/aob/mcr006.

Saseendran, S.A., Ma, L., Nielsen, D.C., Vigil, M.F., Ahuja, L.R., 2005. Simulating planting date effects on corn production using RZWQM and CERES-Maize models. Agron. J. 97, 58–71. Available from: https://doi.org/10.2134/agronj2005.0058.

Schaap, M.G., Leij, F.J., Van Genuchten, M.T., 2001. Rosetta: a computer program for estimating soil hydraulic parameters with hierarchical pedotransfer functions. J. Hydrol. 251, 163–176.

Shapiro, C.A., Wortmann, C.S., 2006. Corn response to nitrogen rate, row spacing, and plant density in Eastern Nebraska. Agron. J. 98, 529–535.

Sharratt, B.S., McWilliams, D.A., 2005. Microclimatic and rooting characteristics of narrow-row versus conventional-row corn. Agron. J. 97, 1129–1135. Available from: https://doi.org/10.2134/agronj2004.0292.

Soil Survey Staff, 2018. Natural Resources Conservation Service, United States Department of Agriculture. Soil Survey Geographic (SSURGO) Database. Available online at https://sdmdataaccess.sc.egov.usda.gov.

Stone, P., Pearson, A., Sorensen, I., Rogers, B., 2000. Effect of row spacing and plant population on maize yield and quality. Agron. N. Z. 30, 67–75.

Subedi, K.D., Ma, B.L., Smith, D.L., 2006. Response of a leafy and non-leafy maize hybrid to population densities and fertilizer nitrogen levels. Crop Sci. 46, 1860–1869. Available from: https://doi.org/10.2135/cropsci2005.06-0141.

Teasdale, J.R., 1998. Influence of corn (*Zea mays*) population and row spacing on corn and velvetleaf (*Abutilon theophrasti*) yield. Weed Sci. 46, 447–453.

Timlin, D., Pachepsky, Y., Acock, B., Šimunek, J., Flerchinger, G., Whisler, F., 2002. Error analysis of soil temperature simulations using measured and estimated hourly weather data with 2DSOIL. Agric. Syst. 72, 215–239.

Timlin, D.J., Fleisher, D.H., Kemanian, A.R., Reddy, V.R., 2014. Plant density and leaf area index effects on the distribution of light transmittance to the soil surface in maize. Agron. J. 106, 1828–1837. Available from: https://doi.org/10.2134/agronj14.0160.

Van Genuchten, M.T., 1980. A closed-form equation for predicting the hydraulic conductivity of unsaturated soils 1. Soil Sci. Soc. Am. J. 44, 892–898.

Van Roekel, R.J., Coulter, J.A., 2012. Agronomic responses of corn hybrids to row width and plant density. Agron. J. 104, 612–620. Available from: https://doi.org/10.2134/agronj2011.0380.

Von Caemmerer, S., 2000. Biochemical Models of Leaf Photosynthesis. Csiro Publishing.

Westgate, M.E., Forcella, F., Reicosky, D.C., Somsen, J., 1997. Rapid canopy closure for maize production in the northern US corn belt: radiation-use efficiency and grain yield. Field Crops Res. 49 (2/3), 249–258. Available from: https://doi.org/10.1016/s0378-4290(96)01055-6.

Yao, A.Y.M., Shaw, R.H., 1964. Effect of plant population and planting pattern of corn on water use and yield. Agron. J. 56, 147–152. Available from: https://doi.org/10.2134/agronj1964.00021962005600020008x.

Yin, X., Goudriaan, J., Lantinga, E.A., Vos, J., Spiertz, H.J., 2003. A flexible sigmoid function of determinate growth. Annal. Bot. 91, 361–371.

Zhu, J., Vos, J., van der Werf, W., van der Putten, P.E.L., Evers, J.B., 2014. Early competition shapes maize whole-plant development in mixed stands. J. Exp. Bot. 65, 641–653. Available from: https://doi.org/10.1093/jxb/ert408.

CHAPTER 4

Soil health benefits of the solar corridor crop system

Robert J. Kremer
University of Missouri, Columbia, MO, United States

Contents

Introduction

Intercropping, the practice of growing two or more crops together in the same row or in adjacent rows close enough for biological interaction, is followed by many farmers around the world to increase overall crop productivity and profitability, improve pest and disease management, and gain better use efficiency of nutrient, water, and light resources (Francis et al., 1986; Khan et al., 2014; Lyon et al., 2009; Vandermeer, 1989). Intercropping systems, including the solar corridor cropping system (SCCS), have been predominantly evaluated for crop production efficiency, yield improvement, and economic benefits for the overall farming enterprise (Deichman, 2009; Khan et al., 2014). The impacts of intercropping on soil properties have received less attention, and most studies assessing various aspects of soil health have occurred in just the last 15–20

The Solar Corridor Crop System
DOI: https://doi.org/10.1016/B978-0-12-814792-4.00004-8

years. The studies to date cover a broad range of intercropping systems in temperate and tropical climates including agroforestry practices, a variety of combinations of cereal, legume, and other food crops, and various planting patterns. Information on effects of the SCCS on soil health maintenance or improvement is limited (Kremer and Deichman, 2014) thereby leading to assumptions that soil health benefits observed in other intercropping practices are also likely to occur in the SCCS. The objectives of this chapter are to provide background on the impacts of intercropping systems on soil health and present the results of recent studies with the SCCS in which various soil parameters were measured as indicators for assessing soil health specifically under this management system.

Soil health

The concept of soil health was proposed by Doran (2002) as the capacity of a living soil to function, within ecosystem boundaries, to sustain plant and animal productivity, maintain or enhance water and air quality, and promote plant and animal health. Coleman et al. (1998) further distinguish soil health from soil quality by suggesting that the health and balanced activity of all groups of organisms within an ecosystem is implicit and should be specifically noted as a component of soil health. Recently Lehman et al. (2015) affirmed the importance of microbial diversity and activity as the basis for soil function by noting that soil health relies on diverse soil biological communities that support high levels of critical environmental services. Kibblewhite et al. (2008) considered soil health within the context of sustainable agriculture to assure that agricultural production does not take priority over provision of ecosystem services. Their working definition was that a "healthy agricultural soil is capable of supporting the production of food and fiber to a level and with quality sufficient to meet human requirements, *and* deliver ecosystem services essential to maintain environmental quality, quality of life for humans, animals, plants and conservation of biodiversity" (Kibblewhite et al., 2008). Soil health functions can be assigned for crop production (i.e., water infiltration capacity, storage/release of nutrients, disease suppression) but functions necessary for important ecosystem services [carbon (C) sequestration, water quality maintenance, biodiversity enhancement] must also be provided as well (Stirling et al., 2016). Optimal soil health requires a balance between soil functions for productivity, environmental quality, and plant and animal health all of which are greatly affected by management and land-use decisions. Good management practices that consider soil health must consider all functions, rather than focus on single functions, such as crop productivity (Doran, 2002). In summary, soil health focuses on the living, dynamic nature of soil that incorporates the biological attributes of biodiversity, soil food web structure, ecosystem functioning, and the intimate relationships of soil microorganisms with plants and animals (Kremer, 2016).

Soil health is an evolving concept based primarily on numerous indicators of chemical and physical properties and relatively few biological indicators for assessment of management impacts on our soil resource. Although many of the crop production functions and nearly all ecosystem services are driven by biological processes (Kibblewhite et al., 2008), microbial diversity and microbial functional groups are not used as standard soil health indicators in assessment models due to a lack of sufficient databases and to difficulty in devising in-field sampling methods that maintain in situ conditions (Lehman et al., 2015). Several soil health assessment models are available to evaluate the effects of land management on the soil resource (Morrow et al., 2016; Stott et al., 2013). However, these models are considered deficient in indicators representing important measures of soil properties such as soil loss (Morrow et al., 2016), disease suppression (van Bruggen and Semenov, 2000), and microbial diversity and biological/biochemical activity (Stott et al., 2013).

Analysis of additional biological indicators, or "bioindicators," is currently used as "supplemental information" with soil health indexes generated via various models. Proposed bioindicators include soil C mineralization, active C (AC or permanganate-oxidizable C), water-extractable (soluble) C, soil enzymes, soil microbial community structure and biodiversity components, soil fauna (i.e., earthworms), and plant disease criteria (Killham and Staddon, 2002; Morrow et al., 2016; Stott et al., 2013; van Bruggen and Semenov, 2000). Furthermore, relationships of soil microorganisms formed by intimate associations with plants strongly suggest that they are major contributors to soil health. Inclusion of reference soils with similar properties aid in gauging potential changes in soil health and associated bioindicators due to different management systems, suggesting that monocropping practices need to be included when assessing impacts of intercropping systems (Sparling, 1997). A functioning microbial community in a healthy soil depends on interactions with properly functioning soil chemical and physical factors (Stirling et al., 2016), thus indicators for these factors along with the bioindicators will be considered in describing impacts of intercropping systems and the SCCS on soil health.

Soil health and sustainable agriculture

Current modern farming practices rely on nonrenewable energy sources, synthetic fertilizers, and pesticides, which often result in soil disturbance or degradation due to tillage and/or bare fallow from postharvest to the next planting season, and degradation of air and water quality. Crop production using these practices likely will not adequately meet food demands; thus sustainable agricultural systems provide alternative practices to protect soils and the surrounding environment in addition to assuring a secure food supply for future generations. Sustainable agriculture involves management practices that conserve soils while maintaining productivity in the long term while simultaneously supporting above- and below-ground biodiversity, which allows continuous agricultural production without harming the broader environment (Stirling et al., 2016). Intercropping with maize (*Zea mays*) may be more sustainable than maize monoculture based on evidence that productivity is higher per unit area of land (Temesgen et al., 2017). Many factors contribute to sustainability under intercropping, which include improved resource capture or higher resource utilization efficiency, and better crop performance under restricted soil moisture and nitrogen levels relative to monoculture. The contribution of nitrogen by a legume component in maize intercropping under low soil nitrogen is an obvious advantage over maize monoculture; however, the complex interactions of all factors that result in overall performance advantages of intercropping are still not well understood (Temesgen et al., 2017). The impact of intercropping on soil health as a critical component in sustainability is even less understood. We aim to demonstrate that intercropping, specifically the SCCS, is a management system that embodies the sustainable agricultural concept with an emphasis on soil health improvement and maintenance.

Intercropping: soil biological activity and soil health

Few studies have established the benefits of intercropping on soil health. Intercropping management systems emphasize improved crop productivity on a land-equivalent basis using sustainable pest management and weed control, and results in additional vegetative production for soil conservation and/or forage for livestock. Soil health improvement is an incidental benefit of intercropping that may be associated with nitrogen fixation by intercropped legumes, increased soil organic matter (SOM), and/or prevention of soil erosion (Khan et al., 2014). Quantitative assessments of soil health indicators to support these benefits are reported for several intercropping practices that vary in types and number of crops in the production system.

Positive impacts of intercropping on soil health have been reported based on responses of various biological and biochemical measurements that were part of large-scale experiments focused on the multiple functions of the production system. For simplicity, the following survey of intercropping impacts on soil health is a general overview compiled from studies using a variety of spatial arrangements of crop mixtures and variable numbers of alternating rows of two or more intercrops. Zhang et al. (2010) measured activities of several nutrient-cycling soil enzymes under 10 different cropping systems and found highest activities with intercropped maize-wheat (*Triticum aestivum*) relative to other rotations or monocropped wheat. In contrast, Wang et al. (2015) reported that soil enzyme activities including urease, nitrate reductase, acid phosphatase, and sucrase in six different intercropping arrangements did not differ from monocropped or rotational counterparts after a 10-year period. The study was conducted on a sandy loam and all treatments received the same fertilizer inputs annually including up to 225 kg N ha^{-1}, which likely masked effects of the legume intercrops on soil fertility buildup and likely repressed soil enzyme activity. Intercropped peanut (*Arachis hypogaea*) and maize increased urease and phosphatase activities relative to monocropped peanut and maize (Chen et al., 2018).

Intercropping also affects the abundance and composition of the soil microbial community. *Rhizobium* species and *Frankia* sp. were associated with nitrogen fixation as well as improved overall soil microbial diversity based on molecular detection methods (Chen et al., 2018). All intercropping combinations of winter wheat, faba bean (*Vicia faba*), and maize increased rhizosphere microbial biomass and significantly altered bacterial community composition compared with monocropping (Song et al., 2007). Populations of earthworms, which are soil health bioindicators that influence both soil structure and primary decomposition, were enhanced in cereal-legume intercrops due to high inputs of organic matter and the absence of tillage relative to conventional monocropped wheat (Schmidt et al., 2003).

Agroforestry systems, intercropping practices that combine high-value tree species planted in multiple rows with agronomic or horticultural crops established in alleyways between the tree rows, play a major role in improving and sustaining soil health (Dollinger and Jose, 2018). In a pecan (*Carya illinoinensis*) agroforestry practice with alleys intercropped to the perennial legume, kura clover (*Trifolium ambiguum*), soil health indicators including soil organic carbon (SOC) content, soil enzyme activities, and stable soil aggregates improved compared with no or unmanaged vegetation in the alleys (Kremer and Kussman, 2011). Similar increases in soil health indicators were observed with other agroforestry systems with managed alley crops (Udawatta et al., 2008, 2009). Arbuscular mycorrhizal fungi (AMF) established symbiotic associations with soybean (*Glycine max*) and barley (*Hordeum vulgare*) that were functionally similar in a Norway spruce (*Picea abies*) agroforestry or monoculture system (Bainard et al., 2013). Mycorrhizae, detected using phospholipid fatty acid (PLFA) analysis of soils in an oak (*Quercus* spp.) agroforestry system, accumulated similar biomass densities in both the grass alleys and within the tree rows but were significantly higher than the adjacent cultivated row-crop field (Unger et al., 2013). Similar trends were found for dehydrogenase and fluorescein diacetate hydrolase, enzymes associated with soil microbial activity and indicative of the size of the microbial population. Soil microbial components (bacteria, fungi, protozoa) were also greater for the tree rows and grass alleys relative to the row-crop field.

The foregoing studies demonstrate that intercropping generally promotes buildup and maintenance of SOC, soil biological activity, and increased microbial diversity that are critical components in improving soil health.

The solar corridor crop system and soil health

Based on previous research showing beneficial effects of various intercropping systems on soil health, we expected the SCCS to have similar impacts on building soil health, an additional benefit to the overall goal of enhancing crop productivity. In contrast to most current intercropping systems that are established on highly productive soils, in organic farming systems, or as the main production practice in subsistence farming, the SCCS will most likely be implemented in fields previously under modern intensive agricultural management on soils that are functionally impaired due to extensive pesticide and inorganic fertilizer inputs, short rotations, tillage, and little, if any, cover cropping. Thus the anticipated soil health-promoting effects of the SCCS are expected to improve sustainability of overall crop production.

The initial SCCS field trial in which we assessed soil health was conducted in 2011 in Knox County, Missouri (40°01′N, 92°11′W), on a Putnam silt loam (fine, montmorillonitic, mesic Vertic Albaqualfs) with two modern maize hybrids: DKC63-42VT3 and DKC62-54VT3. The experiment varied row spacing and seeding rate [76-cm rows at 74,200 seeds ha^{-1}, 152-cm (solar corridor) single- and twin-rows at 37,100 seeds ha^{-1}, and 152-cm (solar corridor) twin-rows at 74,200 seeds ha^{-1}] and was no-till planted. The twin-rows were approximately 19-cm apart, similar to other research (Nelson and Smoot, 2009), but were on 152-cm centers. Soil samples (7.5-cm depth) for soil health assessment were collected midway between maize rows and within maize rows for each row width-maize population combination. The sampling protocol was designed to show the influence of the maize rhizosphere (within row) on soil microbial activity and soil C quality as affected by row width relative to the bulk soil (interrow).

Subsequent field studies conducted over a 4-year period (2013–16) on a Mexico silt loam (fine, smectitic, mesic, Aeric Vertic Epiaqualfs) in Boone County, Missouri (38.89°N, 92.22°W), focused on an expanded set of parameters for both soil C and microbiological bioindicators. We focused on soil C fractions not only because soil C is a key soil health indicator and the key soil property that influences many soil properties (Hatfield et al., 2017), but also because of the anticipated increased release of C substances from roots of maize exposed to more sunlight (Muller et al., 1998) provided by the solar corridor. The more recent studies used single rows planted as a monocrop with 76-cm widths to compare with rows planted at 152-cm widths rather than twin row planting used in the 2011 study. Because the highest soil C fraction contents and microbial activity were found for the in-row soils in 2011 (Kremer and Deichman, 2014), we collected root-soil cores (8.9-cm diameter by 12.5 cm deep) using a bulb planter to sample at the base of maize plants at tassel to silking (VT) to R1 growth stages (Abendroth et al., 2011) in both monocropped and corridor systems. The core samples were then processed to provide root biomass for calculating root density and rhizosphere soil for soil health analyses.

Soil carbon in the solar corridor crop system

We hypothesized that higher photosynthetic fixation of C will increase labile C or "short-lived" C [soluble or water-extractable organic C (WEOC) and active C fractions] released from maize roots and stimulate biological processes in the rhizosphere including root growth and rhizobacterial community diversity and activity. Similar conceptual models for altered C and N allocation to the rhizosphere due to elevated CO_2 also suggested that root growth and microbial community structure likewise are increased (Pregitzer et al., 2007). The input of C substances by roots, known as "rhizodeposition," is a vital energy and substrate source for microbial proliferation and activity in the rhizosphere (Schenck zu Schweinsberg-Mickan et al., 2012). The enhanced rhizosphere microbial community resulting from increased rhizodeposition in SCCS-grown maize may lead to higher N mineralization, synthesis of more plant growth—promoting substances, root growth stimulation, and improved overall plant growth with more efficient nutrient and water use efficiency and stress tolerance.

Total SOC and its component fractions were measured as specific soil health indicators because they are the source of energy that drives the soil food web that sustains abundant microbial communities, high biological activity, and a rich diversity of microbial species to ensure that a full range of soil biological processes and ecosystem services are provided (Bardgett and Wardle, 2010). Rhizosphere analyses show that SOC was relatively stable in maize rows in either monocropped or the solar corridor arrangement. This is shown in Table 4.1 using representative data from the 2016

Table 4.1 Soil organic carbon (SOC), active carbon (AC), and water-extractable carbon (WEOC) contents for corn rhizospheres in monocropped and solar corridor crop systems (SCCS) on a Mexico silt loam.

	SOC		AC		WEOC	
Corn hybrid	**Mono row**[a]	**Corridor row**[b]	**Mono row**	**Corridor row**	**Mono row**	**Corridor row**
	mg kg^{-1} soil				**mg mL^{-1}**	
P1151	16.3	17.8	500	470	36.2	37.4
P23F50	15.0	16.0	490	460	36.5	32.4
MC6150	15.5	17.2	480	485	34.2	34.7
MC6580	15.2	16.6	500	520	33.0	35.1
MC527	16.8	18.2	470	510	31.6	37.2
A6866	17.5	18.0	500	490	28.0	35.8
A7664	16.4	17.0	485	530	27.0	34.8
A5565	15.8	16.6	520	520	30.9	33.0

Soils were sampled at V10 corn growth developmental stage during the 2016 growing season. Maize hybrids: Pioneer P1151, Pioneer P23F50, Masters Choice 6150, Masters Choice 6580, Masters Choice 527, Augusta 6866, Augusta 7664, Augusta 5565.

[a]Mono row, in-row soil sample taken in monocropped corn planted in 76-cm rows.

[b]Corridor row, in-row soil sample taken in row adjacent to the 152-cm wide corridor.

study where SOC ranged from 15.03 to 18.23 mg kg^{-1} soil across all maize hybrids planted in either cropping system. The slight increases noted for some of the hybrids in the corridor row relative to the monocropped rows might result from higher root densities attained within the SCCS arrangement (Fig. 4.1) (Kremer and Deichman, 2014). A more pronounced impact of planting arrangement on soil C dynamics was noted for active C, which increased by up to 10% in 4 of 10 hybrids when planted within the SCCS (Table 4.1). A similar effect of the SCCS was observed for WEOC, which increased by 25% or more in 5 of the 10 hybrids (Fig. 4.2). Interestingly all hybrids contributing to increases in both labile C pools were selections incorporating "older genetics" in nontransgenetic lines, and responded in increased or similar grain yields when planted in the SCCS (see Chapter 2: Variety selection for the solar corridor crop system (SCCS)). Labile C fractions are generated through maintaining soil biological activity and are enhanced under sustainable and regenerative production systems, contributing to higher maize yields and reduced inputs of fertilizers and herbicides and ultimately to higher profitability relative to conventional production systems (LaCanne and Lundgren, 2018).

Certain hybrids, primarily Masters Choice and Augusta, likely possess lineage from early developed hybrids that are able to respond positively to changes in environment, or flex, which include altered row widths, changes in plant density, increased solar radiation, and potentially

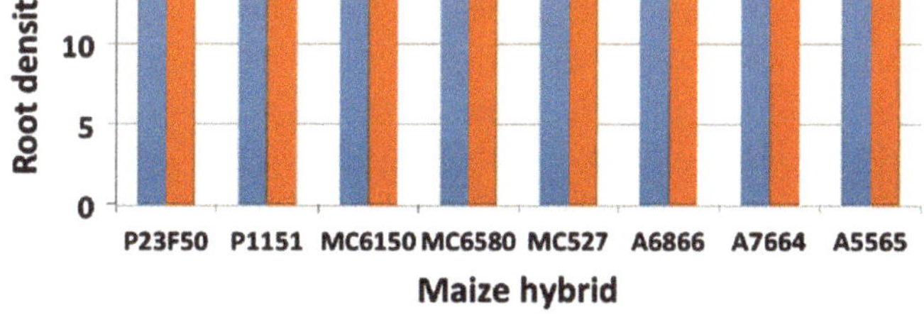

Figure 4.1 Root density of corn hybrids planted as a monocrop or in the solar corridor crop system on a Mexico silt loam. Soils were sampled at VT corn growth developmental stage during the 2014 growing season. Maize hybrids: Pioneer P23F50, Pioneer P1151, Masters Choice 6150, Masters Choice 6580, Masters Choice 527, Augusta 6866, Augusta 7664, Augusta 5565.

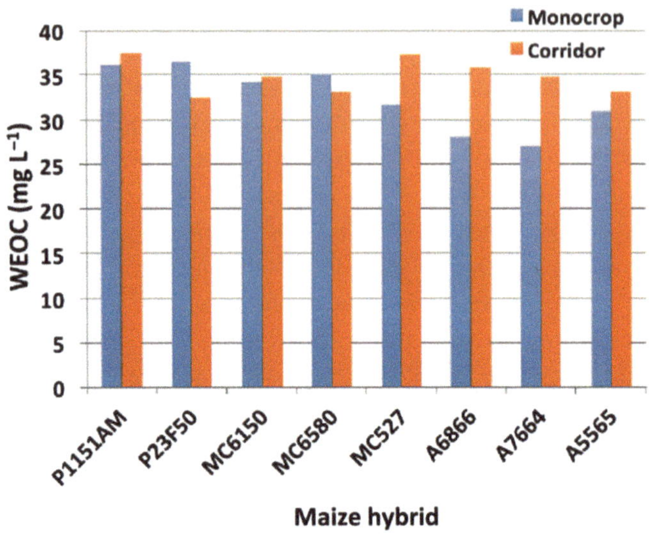

Figure 4.2 Water-extractable organic C (WEOC) from rhizospheres of corn hybrids planted as a monocrop or in the solar corridor crop system on a Mexico silt loam. Soils were sampled at VT corn growth developmental stage during the 2016 growing season. Maize hybrids: Pioneer P23F50, Pioneer P1151, Masters Choice 6150, Masters Choice 6580, Masters Choice 527, Augusta 6866, Augusta 7664, Augusta 5565.

increased CO_2 availability, provided in these studies. Pioneer 1151, included in the 2016 trials, is one of the very few "modern" hybrids that responded in the SCCS, which may be a result of selection for high drought tolerance (Luce, 2016, personal communication). Results of field trials to date suggest that the SCCS provides an environment for favorable light intensity, maximum CO_2 capture, and increased root density that are all factors contributing to the release of greater amounts of C substances from maize roots (Paterson, 2003). Our results confirmed that labile C fractions are sensitive indicators of soil health (Culman et al., 2012; Veum et al., 2012) because changes were detected soon, that is, within the same growing season during establishment of the SCCS (Fig. 4.1). This contrasts with the gradual changes in total SOC, which may require several growing seasons to detect (Miles and Brown, 2011).

Higher labile C accumulating in the maize rhizosphere is believed to be related to increased photosynthetic activity of maize exposed to more solar radiation provided in the SCCS. Increased photosynthesis can be detected indirectly by nondestructively measuring the chlorophyll content of intact leaves with mature chloroplasts. A soil and plant analyzer development (SPAD) chlorophyll meter was used to detect differences in far-red wavelengths related to leaf chlorophyll concentration, which is

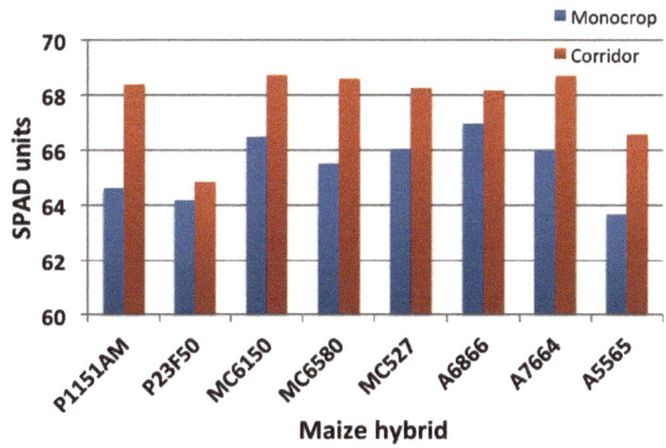

Figure 4.3 SPAD units of leaf area of the ear leaf of corn hybrids planted as a monocrop or in the solar corridor crop system on a Mexico silt loam. Leaves were monitored at VT corn growth developmental stage during the 2014 growing season. Maize hybrids: Pioneer P23F50, Pioneer P1151, Masters Choice 6150, Masters Choice 6580, Masters Choice 527, Augusta 6866, Augusta 7664, Augusta 5565.

expressed in dimensionless "SPAD units" ranging from 0 to 80. SPAD units are highly correlated with chlorophyll content per unit leaf area and also indicate responses to environmental stresses including reduced light (Naus et al., 2010; Xiong et al., 2015). Our intent was to detect effects of increased light on leaf chlorophyll content and compare with rhizosphere labile C content to demonstrate the potential for increased photosynthetically fixed C by maize in the SCCS. Field measurements taken on maize ear leaves (stage VT) showed from 1% to 6.35% higher SPAD readings for all hybrids grown in the SCCS relative to those grown in the monocropped system (Fig. 4.3). Interestingly a modest correlation ($r^2 = 0.52$) between WEOC and SPAD units suggest that higher soluble carbohydrate contents released by maize roots was likely related to higher chlorophyll and, indirectly, higher photosynthesis that provided more fixed CO_2 in maize growing in the SCCS. This agrees with previous reports regarding increased carbohydrate contents released by growing maize roots in response to higher levels of C fixation (Muller et al., 1998).

The soil microbiome

The soil microbiome includes groups of different microorganisms inhabiting particular niches within the soil, their "biomes," and is used interchangeably with "soil microbial community." We measured the maize rhizosphere soil microbial community using PLFA analysis. This analysis detects lipids of microbial membranes as "biomarkers" for specific groups of microorganisms producing a profile or "fingerprint" of the community structure. Biomarkers specific to functional groups of microorganisms include bacteria, actinobacteria, AMF, saprophytic or non-AM fungi, and protists. Thus rapid changes in the microbial community structure are based on changes in PLFA patterns, which we expected to occur in the rhizospheres of maize in the SCCS. Also, the total PLFA content is a measure of viable microbial biomass (Zelles, 1999). Based on studies over a 4-year period, rhizospheres of specific maize hybrids grown in the SCCS supported higher total microbial biomass compared with hybrids grown in the monocropped system. Total PLFA is shown as the cumulative bar height for each maize hybrid in Fig. 4.4. Although total PLFA does not provide information on microbial community composition, higher biomass suggests a more complex community structure relative to rhizospheres with lower total PLFA. Microbial biomass is constantly recycled through a cycle of biogenesis from active growth to decomposition (microbial turnover) eventually contributing to organic matter formation and N mineralization, important for soil health and plant nutrition, respectively. Thus increased microbial biomass resulting from increased rhizodeposition of labile C contents in the maize root zone is indicated by higher total PLFA, which supplements maize productivity in the SCCS, similar to that suggested by LaCanne and Lundgren (2018) for regenerative agriculture systems.

Similar to total biomass PLFA, SCCS-grown maize showed more robust rhizosphere microbial community composition as illustrated by the 2014 field trial results (Fig. 4.4). Total soil bacteria were increased for nearly all hybrids; however, Gram-negative bacteria were particularly enhanced by three of four Masters Choice hybrids and by Augusta 6866—these hybrids were selected based on their expected response to increased solar radiation in the SCCS. Gram-negative bacteria rapidly metabolize readily available C—in this case, in the forms of WEOC and AC—released via rhizodeposition that support the development of more diverse microbial communities (Bardgett and Wardle, 2010). Many Gram-negative rhizobacteria are involved in nutrient cycling, including N mineralization, produce plant growth–regulating compounds that stimulate root growth, and antibiotics that suppress soil and root pathogens. Higher densities of AMF associated with roots of selected hybrids suggest that increased carbohydrate content released by SCCS-grown maize

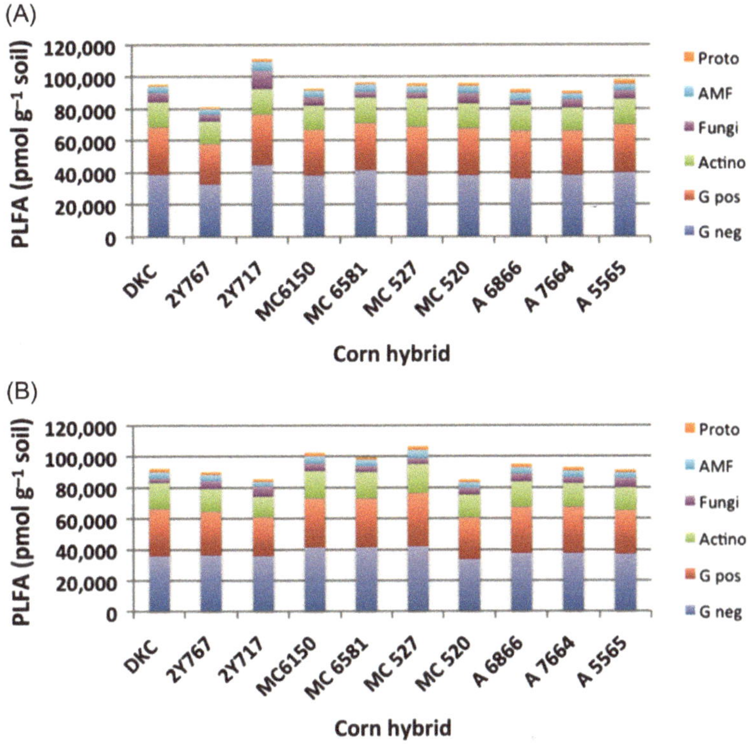

Figure 4.4 Phospholipid fatty acid profiles (PLFA) of the major microbial components in rhizosphere soils of corn produced as a monocrop or in a solar corridor system on Mexico silt loam. Soil collected in-row from both production systems at V10 corn growth developmental stage during the 2014 growing season. Maize hybrids: Dekalb DKC62-97, Mycogen 2Y767, Mycogen 2Y717, Masters Choice 6150, Masters Choice 6581, Masters Choice 527, Masters Choice 520, Augusta 6866, Augusta 7664, Augusta 5565. (A) Monocropped maize; (B) solar corridor maize.

supported effective symbioses for exchanging plant-available nutrients and water extracted from root-free soil by AMF hyphae.

Our observations based on limited field trials are that maize hybrids possessing traits incorporated from older genetic lines produced denser root systems and released higher labile C contents thereby supporting a more diverse root microbiome. These conclusion parallel those of Schmidt et al. (2016) that the current directed selection pressure of modern maize hybrids for yield and above-ground traits under high-input, high-density environments inadvertently alter root system traits and ecophysiological functioning. As a consequence, this selection pressure likely altered root exudate quantity and quality and affected coevolution of the maize rhizosphere microbiome, which altered nutrient acquisition and other growth promotion functions. Thus investigating factors contributing to inadvertent belowground changes during artificial selection of hybrids may help maize breeders develop hybrids with the ability to form adequate root systems to support beneficial microbial interactions under soil health-building and sustainable management practices (Schmidt et al., 2016).

Soil microbial activity

Soil glucosidase activity is a sensitive soil health indicator for microbial activity and is used to differentiate various soil and crop management regimes (Stott et al., 2010). Glucosidase activity was highest in the SCCS for the majority of hybrids across all seasons (Kremer et al., 2015a). The impact of the SCCS likely reflects the higher densities of fine roots, greater amounts of rhizodeposited substrates, and diverse microbial communities (Liang et al., 2002; Paterson, 2003) relative to monocropped maize. In general the Masters Choice and Augusta hybrids tended to promote soil glucosidase activity; however, the magnitude of activity within the SCCS was inconsistent across seasons and hybrids and strong positive correlations with other soil health parameters were not observed.

Although soil glucosidase activity is considered a good indicator of management-induced changes in soil health (Bandick and Dick, 1999; Stott et al., 2010), it may not be sensitive enough for differentiating effects of cropping system and specific maize hybrid on microbial activity in the rhizosphere. The substrate for glucosidase, cellobiose, is a product of cellulose decomposition, yet about 80% of maize root exudates are water-soluble carbohydrates, organic acids, and amino acids (Hütsch et al., 2002). Because of the wide C:N of these root exudates, rapid metabolism by the proliferating rhizosphere microbial community results in temporary N immobilization before synthesis of extracellular enzymes is adequate for mineralizing N from organic material and dead microbial biomass (Paterson, 2003; Schenck zu Schweinsberg-Mickan et al., 2012).. Thus more sensitive soil health indicators for microbial activity should include enzymes that degrade soluble carbohydrates such as sucrose or glucose and/or amino acids present in the rhizosphere. These soil enzymes will be examined in future soil health assessments of the SCCS.

Intercropping in the solar corridor crop system: impacts of floor crops on soil health

An important attribute of the SCCS design is the integration of low stature crops, or "floor crops," planted into the corridor to intercept sunlight that reaches the soil surface between the maize rows bordering the corridor (Fig. 4.5). Floor crops strengthen the sustainability of the SCCS by increasing plant diversity, improving soil conservation with surface soil coverage, reducing or eliminating interrow cultivation, enhancing SOM with inputs of root biomass and vegetative residues with narrower C:N to supplement the accumulated maize residues, providing additional microbial diversity, and improving overall soil health. Previous intercropping studies confirm the provision of these beneficial properties by cover crops interplanted in various crops such as tomato that promoted soil microbial biomass and biodiversity (Buyer et al., 2010). Cover crops also stimulate mycorrhizal communities that associate with various subsequent cash crops to enhance nutrient acquisition, disease resistance, and drought tolerance (Lehman et al., 2012). Thus we can expect that actively growing cover crops in the corridor during the fallow period after maize harvest would provide a continuous source of C and energy for the microbial

Figure 4.5 The solar corridor crop system planted with maize (R) next to monocropped maize in 76-cm row widths (L) on Mexico silt loam soil in Boone County, Missouri, USA. The corridor is 152 cm wide and is planted with cowpea. Maize is at the dent stage (reproductive growth stage R5) and cowpea is at pod-filling stage. *Photo: R.J. Kremer.*

community, as shown for cover crops in agroforestry practices (Cardinael et al., 2018). Our preliminary SCCS study likewise suggested that winter wheat sustained important soil processes including nutrient cycling, SOC buildup, and soil aggregate formation when included as a winter cover crop (Kremer and Deichman, 2014).

More recent field studies have focused primarily on cowpea (*Vigna unguiculata*) because of its N-fixing ability and tolerance to shade and moisture stress (Fig. 4.5). Cowpea increased labile soil C in the corridor by 25% compared to monoculture and released 20% higher WEOC into the soil compared with SCCS-grown maize hybrids (Kremer et al., 2017). Cowpea yielded over 600 kg ha^{-1} of vegetative biomass for cycling into SOC and, at 4% N content, added about 24 kg of organic N per ha that is available for microbial transformation and plant uptake. Although results with cowpea as a floor crop show that it provides abundant vegetative residues for enhancing SOC and the soil microbial community, other crops may also readily adapt to SCCS conditions and should be evaluated as possible floor crop candidates. For example, recent studies showing that over 90% of N in field pea residues originates from N fixation (Knight, 2012) suggest that pea as a floor crop could be a valuable source of organic N and aid in reducing synthetic N inputs in the SCCS. A variety of cover crops may be used as floor crops to not only maintain soil health but offer other uses including weed management while minimizing herbicide inputs (Davis et al., 2012). Also, a preliminary evaluation of several floor crops for multiple benefits in 2015 revealed that certain crops such as buckwheat (*Fagopyrum esculentum*) provide ecosystem services such as attraction of beneficial arthropods including pollinating and predatory species (Kremer et al., 2015b).

Conclusion

We show how the SCCS provides an agronomic design for improving soil health that is synchronized with its principal role of efficiently capturing C through photosynthesis to enhance maize productivity. The higher photosynthetic activity that leads to increased labile C contents released into soil through maize roots seems to be the primary means responsible for enhancing soil health. The SCCS can be a management practice within a sustainable agriculture framework that not only improves production efficiency of a crop but also SOM quality, the structural and functional characteristics of the soil biological community, and the balance

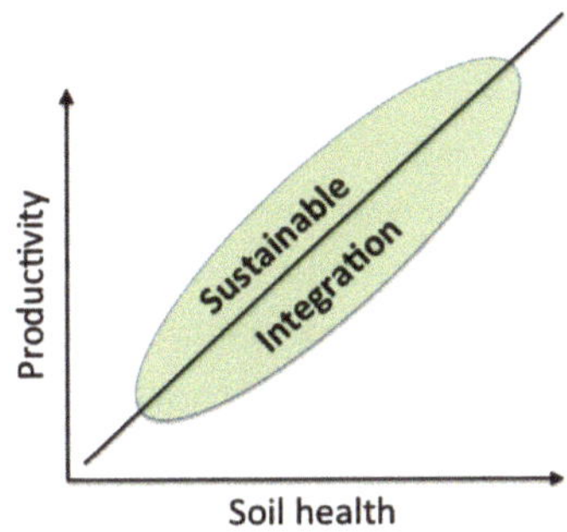

Figure 4.6 Conceptual model for the interaction of increased productivity potential provided by the solar corridor crop system with increasing integration of sustainable practices relative to soil health. *Modified from the concept presented in Hatfield, J.L., Sauer, T.J., Cruse, R.M., 2017. Soil: the forgotten piece of the water, food, energy nexus. Adv. Agron. 143, 1–46.*

that sustains the broader environment, that is, "ecosystem services" (Stirling et al., 2016).

To achieve sustainable production the SCCS will require selected agronomic inputs and practices that function optimally and are compatible for both agronomic production and soil health (Hatfield et al., 2017). Thus a number of sustainable practices can be integrated within the SCCS to increase productivity and soil health, shown conceptually in Fig. 4.6. Sustainable practices for integration may include crop rotation, conservation tillage, residue retention, and organic matter buildup within the SCCS to improve SOC, soil structure stability, and soil water retention as effective strategies in restoring soil health and productivity in soils degraded by tillage and other intensive practices (Hatfield, 2014). Integration of these practices are known to enhance or restore soil microbial diversity and functional activity to mediate nutrient release from organic sources while potentially reducing pesticide and inorganic fertilizer inputs (Jangid et al., 2008; Garcia-Orenes et al., 2013).

Optimal productivity of the SCCS depends on two primary factors. First is selection of specific maize hybrids for high-yield performance in response to SCCS conditions as the foundation. However, screening of both older and nontransgenic as well as modern hybrids that respond with high yields and maintain root and rhizosphere function (Schmidt et al., 2016) under SCCS conditions must continue for this system to be operational in conventional and sustainable crop production settings on various soils in different geographic regions. Indeed, selected maize hybrids evaluated over a 4-year period (2013–16) produced higher or similar grain yields relative to monocropped yields (see Chapter 2: Variety selection for the solar corridor crop system (SCCS)) and simultaneously contributed to higher labile soil C, microbial activity, and microbial biomass and diversity (Kremer et al., 2015a). The second factor is compatible floor crops for the corridor to intercept sunlight reaching the soil surface benefits soil health and offer options to the farmer for cover cropping, livestock grazing, or as harvestable grain, food or seed crops. More screening is necessary to identify floor crops best suited for adequate yields and for maintaining or improving soil properties in the overall SCCS. Results with the limited number of maize hybrids and floor crops to date confirm that deliberate management of crops and soils, as implemented with the SCCS, has a considerable impact on soil biology and soil health (Bossio et al., 2005).

Research quantifying effects of cropping systems similar to the SCCS on soil health is extremely limited. An integrated maize cropping system with wide row spacing, no-tillage, and residue retention in northeast China revealed that microbial biomass increased by 30%, and selected soil enzyme activities increased by over 80% compared to maize monocropping in standard row widths (Sun et al., 2018). Implementation of a "modified SCCS" on some Iowa farms using cover crop mixtures as floor crops in 152-cm corridors qualitatively improved soil health and yielded similarly to monocropped maize in 76-cm rows (Houghton, 2018). Similar observations were found with maize and sunflower (*Helianthus annuus*) in 152-cm rows in on-farm production fields in South Dakota (Mullally, 2011). The positive responses reported for these trials show the potential of the SCCS for improving soil health in diverse landscapes. As the SCCS develops based on needs presented in this book, farmers will be able to adapt the SCCS approach for their crop production system and, by closely following the management practices required to fully exploit the benefits of the system, realize consistent and economic yields as well as improved soil health and ecosystem services.

Research needs for the solar corridor crop system as a soil health improvement practice

Concerns about effects of current industrial agricultural production systems on degradation of soils and landscapes, biodiversity loss, and reduced environmental quality have led to development of alternative models of farm production that promote soil health and biodiversity (LaCanne and Lundgren, 2018). Our initial studies have demonstrated the potential of the SCCS as a model system for improving soil health that could be a practice that farmers adopt to restore soil health and productivity on their land. However, more research is required to fully document the numerous benefits to soil health that are possible with implementation of the SCCS. Major research needs are as follows:

- Quantitative evaluation of CO_2 fixation, and photosynthetic activity to better document carbon dynamics from the plant to rhizosphere to soil microbiome.
- More sensitive evaluation of the rhizosphere microbiome using techniques such as metagenomics (Rathore et al., 2017) to better describe the SCCS maize–microbiome relationship with healthy crops and soils.
- Evaluation of specific functional microbial groups affected by increased labile C inputs including mycorrhizal fungi and plant growth–regulating rhizosphere microorganisms (Song et al., 2007).
- Expand soil health bioindicator measurements to include relevant soil enzyme assays, C metabolism including soil/microbial respiration, and N transformations.
- Replicated studies of the SCCS paired with conventional, sustainable, organic, and subsistence systems across various crop production regions to validate the protocol and soil health assessment.
- Implement a more systematic experimental approach to include rotation of maize into floor crop strips in the same field over a series of at least 3 years in order to track soil health restoration effects over time.
- Investigate effects of practical variations in agronomic practices on soil health within the SCCS including no-tillage, various uses of floor crops, other "tall stature" crops besides maize, and integration of livestock.

References

Abendroth, L.J., Elmore, R.W., Boyer, M.J., Marlay, S.K., 2011. Corn Growth and Development, PMR 1009. Iowa State University Extension, Ames, IA.

Bainard, L.D., Koch, A.M., Gordon, A.M., Klironomos, J.N., 2013. Growth response of crops to soil microbial communities from conventional monocropping and tree-based intercropping systems. Plant Soil 363, 345–356.

Bandick, A.K., Dick, R.P., 1999. Field management effects on soil enzyme activities. Soil Biol. Biochem. 31, 1471–1479.

Bardgett, R.D., Wardle, D.A., 2010. Aboveground-Belowground Linkages: Biotic Interactions, Ecosystem Processes, and Global Change. Oxford University Press, Oxford.

Bossio, D.A., Girvan, M.S., Verchot, I., Bullimore, J., Borelli, T., Albrecht, A., et al., 2005. Soil microbial community response to land use changes in an agricultural landscape of western Kenya. Microb. Ecol. 49, 50–62.

Buyer, J.S., Teasdale, J.R., Roberts, D.P., Zasada, I.A., Maul, J.E., 2010. Factors affecting soil microbial community structure in tomato cropping systems. Soil Biol. Biochem. 42, 831–841.

Cardinael, R., Guenet, B., Chevallier, T., Dupraz, C., Cozzi, T., Chenu, C., 2018. High organic inputs explain shallow and deep SOC storage in long-term agroforestry system–combining experimental and modeling approaches. Biogeosciences 15, 297–317.

Chen, J., Arafat, Y., Wu, L., Xiao, Z., Li, Q., Khan, M.A., et al., 2018. Shifts in soil microbial community, soil enzymes and crop yield under peanut/maize intercropping with reduced nitrogen levels. Appl. Soil Ecol. Available from: https://doi.org/10.1016/j.apsoil.2017.11.010.

Coleman, D.C., Hendrix, P.F., Odum, E.P., 1998. Ecosystem health: an overview. In: Huang, P.M. (Ed.), Soil Chemistry and Ecosystem Health. SSSA Spec. Publ. 52. Soil Science Society of America, Madison, WI, pp. 1–20.

Culman, S.W., Snapp, S.S., Freeman, M.A., Schipanski, M.E., Beniston, J., Lal, R., et al., 2012. Permanganate oxidizable carbon reflects a processed soil fraction that is sensitive to management. Soil Sci. Soc. Am. J. 76, 494–504.

Davis, A.S., Hill, J.D., Chase, C.A., Johanns, A.M., Liebman, M., 2012. Increasing cropping system diversity balances productivity, profitability and environmental health. PLoS One 7 (10), e47149. Available from: https://doi.org/10.1371/journal.pone.0047149.

Deichman, C. LeRoy, 2009. A sustainable farming system to maximize photosynthesis. In: Proc. Farm. Syst. Design Symp., Monterey, CA, pp. 87–88.

Dollinger, J., Jose, S., 2018. Agroforestry for soil health. Agroforest. Syst. 92, 213–219.

Doran, J.W., 2002. Soil health and global sustainability: translating science into practice. Agric. Ecosyst. Environ. 88, 119–127.

Francis, C., 1986. Multiple Cropping Systems. Macmillan Publishers, New York.

Garcia-Orenes, F., Morugán-Coronado, A., Zornoza, R., Scow, K., 2013. Changes in soil microbial community structure influenced by agricultural management practices in a Mediterranean agro-ecosystem. PLoS One 8 (11), e80522. Available from: https://doi.org/10.1371/journal.pone.0080522.

Hatfield, J.L., 2014. Soil degradation, land use, and sustainability. In: Songstad, D.D., Hatfield, J.L., Tomes, D.T. (Eds.), Convergence of Food Security, Energy Security and Sustainable Agriculture. Springer-Verlag, Berlin, pp. 61–74.

Hatfield, J.L., Sauer, T.J., Cruse, R.M., 2017. Soil: the forgotten piece of the water, food, energy nexus. Adv. Agron. 143, 1–46.

Houghton, D., 2018. A walk on the wide side. *The Furrow*. April issue, <https://www.johndeerefurrow.com/2018/04/05/awalk-on-the-wide-side/>.

Hütsch, B.W., Augustin, J., Merbach, W., 2002. Plant rhizodeposition – an important source for carbon turnover in soils. J. Plant Nutr. Soil Sci. 165, 397–407.

Jangid, K., Williams, M.A., Franzluebbers, A.J., Sanderlin, J.S., Reeves, J.H., et al., 2008. Relative impacts of land-use, management intensity and fertilization upon soil microbial community structure in agricultural systems. Soil Biol. Biochem. 40, 2843–2853.

Khan, Z.R., Midega, C.A.O., Pittchar, J.O., Pickett, J.A., 2014. Push-pull: a novel IPM strategy for the green revolution in Africa. In: Peshin, R., Pimentel, D. (Eds.), Integrated Pest Management. Springer-Verlag, Dordrecht, pp. 333–348.

Kibblewhite, M.G., Ritz, K., Swift, M.J., 2008. Soil health in agricultural systems. Philos. Trans. R. Soc. Biol. Sci. 363, 685–701.

Killham, K., Staddon, W.J., 2002. Bioindicators and sensors of soil health and the application of geostatistics. In: Burns, R.G., Dick, R.P. (Eds.), Enzymes in the Environment: Activity, Ecology, and Applications. Marcel Dekker, New York, pp. 391–405.

Knight, J.D., 2012. Frequency of field pea in rotations impacts biological nitrogen fixation. Can. J. Plant Sci. 92, 1005–1011.

Kremer, R.J., 2016. Soil health, quality & microbial diversity. Acres U.S.A. 46 (12), 38–43.

Kremer, R.J., Deichman, C. LeRoy, 2014. Soil quality and the solar corridor crop system. Agron. J. 106, 1853–1858.

Kremer, R.J., Kussman, R.D., 2011. Soil quality in a pecan-kura clover alley cropping system in the Midwestern USA. Agroforest. Syst. 83, 213–223.

Kremer, R.J., Deichman, C. LeRoy, Reinbott, T.M., 2017. Multiple benefits of intercropping cowpea in the solar corridor cropping system. In: Agronomy Abstracts. ASA-CSSA-SSSA Annual Meetings, Tampa, FL.

Kremer, R.J., Deichman, C.LeRoy, Reinbott, T.M., Veum, K.S., 2015a. Soil biology is enhanced through synergistic relationships in the solar corridor production system. In: Agronomy Abstracts. ASA-CSSA-SSSA Annual Meetings, Minneapolis, MN.

Kremer, R.J., Reinbott, T.M., Veum, K.S., Deichman, C.LeRoy., Woods, T.L., et al., 2015b. A novel integrated cropping system for efficient grain production, improved soil quality and enhanced beneficial arthropod communities. In: Agronomy Abstracts. ASA-CSSA-SSSA Annual Meetings, Minneapolis, MN.

LaCanne, C.E., Lundgren, J.G., 2018. Regenerative agriculture: merging farming and natural resource conservation profitably. Peer J. 6, e4428. Available from: https://doi.org/10.7717/peerj.4428.

Lehman, R.M., Taheri, W.I., Osborne, S.L., Buyer, J.S., Douds Jr, D.D., 2012. Fall cover cropping can increase arbuscular mycorrhizae in soils supporting intensive agricultural production. Appl. Soil Ecol. 61, 300–304.

Lehman, R.M., Acosta-Martinez, V., Buyer, J.S., Cambardella, C.A., Collins, H.P., Ducey, T.F., et al., 2015. Soil biology for resilient, healthy soil. J. Soil Water Conservat. 70, 12A–18A.

Liang, B.C., Wang, X.L., Ma, B.L., 2002. Maize root-induced change in soil organic pools. Soil Sci. Soc. Am. J. 66, 845–847.

Lyon, D.J., Pavlista, A.D., Hergert, G.W., Klein, R.N., Shapiro, C.A., Knezevic, S., et al., 2009. Skip-row planting patterns stabilize corn grain yields in the central Great Plains. Crop Manag. Available from: https://doi.org/10.1094/CM-2009-0224-02-RS <www.plantmanagementnetwork.org/cm/>.

Miles, R.J., Brown, J.R., 2011. The sanborn field experiment: implications for long-term soil organic carbon levels. Agron. J. 103, 268–278.

Morrow, J.G., Huggins, D.R., Carpenter-Boggs, L.A., Reganold, J.P., 2016. Evaluating measures to assess soil health in long-term agroecosystem trials. Soil Sci. Soc. Am. J. 80, 450–462.

Mullally, S., 2011. Planting in 60″ rows. Sunflower 37 (6), 1–5.

Muller, B., Stosser, M., Tardieu, F., 1998. Spatial distributions of tissue expansion and cell division rates are related to irradiance and to sugar content in the growing zone of maize roots. Plant Cell Environ. 21, 149–158.

Naus, J., Prokopova, J., Rebicek, J., Spundova, M., 2010. SPAD chlorophyll meter reading can be pronouncedly affected by chloroplast movement. Photosynth. Res. 105, 265–271.

Nelson, K.A., Smoot, R.L., 2009. Twin- and single-row corn production in Northeast Missouri. Crop Manag. Available from: https://doi.org/10.1094/CM-2009-0130-01-RS<www.plantmanagementnetwork.org/cm/>.

Paterson, E., 2003. Importance of rhizodeposition in the coupling of plant and microbial productivity. Eur. J. Soil Sci. 54, 741–750.

Pregitzer, K.S., Zak, D.R., Loya, W.M., Karberg, N.J., King, J.S., Burton, A.J., 2007. The contribution of root-rhizosphere interactions to biogeochemical cycles in a changing world. In: Cardon, A.G., Whitbeck, J.G. (Eds.), The Rhizosphere, An Ecological Perspective. Elsevier Academic Press, San Diego, pp. 155–178.

Rathore, R., Dowling, D.N., Forristal, P.D., Spink, J., Cotter, P.D., Bulgarelli, D., et al., 2017. Crop establishment practices are a driver of the plant microbiota in winter oilseed rape (*Brassica napus*). Front. Microbiol 8, 1489. Available from: https://doi.org/10.3389/fmicb.2017.01489.

Schenck zu Schweinsberg-Mickan, M., Jörgensen, R.G., Müller, T., 2012. Rhizodeposition: its contribution to microbial growth and carbon and nitrogen turnover within the rhizosphere. J. Plant Nutr. Soil Sci. 175, 750–760.

Schmidt, J.E., Bowles, T.M., Gaudin, A.C.M., 2016. Using ancient traits to convert soil health into crop yield: impact of selection on maize root and rhizosphere function. Front. Plant Sci. 7, 373. Available from: http://dx.doi.103389/fpls.2016.00373.

Schmidt, O., Clements, R.O., Donaldson, G., 2003. Why do cereal-legume intercrops support large earthworm populations?. Appl. Soil Ecol. 22, 181–190.

Song, Y.N., Zhang, F.S., Marschner, P., Fan, F.L., Gao, H.M., Bao, X.G., et al., 2007. Effect of intercropping on crop yield and chemical and microbiological properties in rhizosphere of wheat (*Triticum aestivum* L.), maize (*Zea mays* L.), and faba bean (*Vicia faba* L.). Biol. Fert. Soils 43, 565–574.

Sparling, G.P., 1997. Soil microbial biomass, activity and nutrient cycling as indicators of soil health. In: Pankhurst, C., Doube, B.M., Gupta, V.V.S.R. (Eds.), Biological Indicators of Soil Health. CABI, Wallingford, pp. 97–120.

Stirling, G., Hayden, H., Pattison, T., Stirling, M., 2016. Soil health, soil biology. Soilborne Diseases and Sustainable Agriculture. CSIRO Publishing, Clayton South.

Stott, D.E., Andrews, S.S., Liebig, M.A., Wienhold, B.J., Karlen, D.L., 2010. Evaluation of β-glucosidase activity as a soil quality indicator for the soil management assessment framework. Soil Sci. Soc. Am. J. 74, 107–119.

Stott, D.E., Cambardella, C.A., Karlen, D.L., Harmel, R.D., 2013. A soil quality and metabolic activity assessment after fifty-seven years of agricultural management. Soil Sci. Soc. Am. J. 77, 903–913.

Sun, L., Song, F., Liu, S., Cao, Q., Liu, F., Zhu, X., 2018. Integrated agricultural management practice improves soil quality in Northeast China. Arch. Agron. Soil Sci. Available from: https://doi.org/10.1080/03650340.2018.1468077.

Temesgen, A., Fakai, S., Rodriguez, D., 2017. Intercropping in sustainable maize cultivation. In: Watson, D. (Ed.), Achieving Sustainable Cultivation of Maize, Volume 2, Cultivation Techniques, Pest and Disease Control. Burleigh Dodds Science Publishing Ltd, Cambridge, pp. 203–225.

Udawatta, R.P., Kremer, R.J., Adamson, B.W., Anderson, S.H., 2008. Variations in soil aggregate stability and enzyme activities in a temperate agroforestry practice. Appl. Soil Ecol. 39, 153–160.

Udawatta, R.P., Kremer, R.J., Garrett, H.E., Anderson, S.H., 2009. Agroforestry buffer effects on soil physical properties and enzyme activity on a row-cropped watershed. Agric. Ecosyst. Environ. 131, 98–104.

Unger, I.M., Goyne, K.W., Kremer, R.J., Kennedy, A.C., 2013. Microbial community diversity in agroforestry and grass vegetative filter strips. Agroforest. Syst. 87, 395–402.

van Bruggen, A.H.C., Semenov, A.M., 2000. In search of biological indicators for soil health and disease suppression. Appl. Soil Ecol. 15, 13–24.

Vandermeer, J., 1989. The Ecology of Intercropping. Cambridge University Press, Cambridge.
Veum, K.S., Goyne, K.W., Kremer, R.J., Motavalli, P.P., 2012. Water-stable aggregates and soil organic matter fractions under agricultural and conservation management practices. Soil Sci. Soc. Am. J. 76, 2143–2153.
Wang, Z., Bao, X., Li, X., Jin, X., Zhao, J., Sun, J., et al., 2015. Intercropping maintains soil fertility in terms of chemical properties and enzyme activities on a timescale of one decade. Plant Soil 391, 265–282.
Xiong, D., Chen, J., Yu, T., Gao, W., Ling, X., Li, Y., et al., 2015. SPAD-based leaf nitrogen estimation is impacted by environmental factors and crop leaf characteristics. Sci. Rep. 5, 13389. Available from: https://doi.org/10.1038/srep13389.
Zelles, L., 1999. Fatty acid patterns of phospholipids and lipopolysaccharides in the characterization of microbial communities in soil: a review. Biol. Fert. Soils 29, 111–129.
Zhang, Y.L., Chen, L.T., Sun, C.S., Wu, Z.T., Chen, Z.H., Dong, G.H., 2010. Soil hydrolase activities and kinetic properties as affected by wheat cropping systems of northeastern China. Plant Soil Environ. 56, 526–532.

CHAPTER 5

Putting it all together

C. LeRoy Deichman
Deichman Consulting, Shelbyville, MO, United States

Contents

By this point in the book we have offered the physiological and broad-based (bio)technical basics and basis of how the solar corridor (SC) light-enriched production environment could enable the nearly inconceivable increases in hybrid-specific corn yields we achieved, while simultaneously improving our vital and finite soil resources. This chapter endeavors to capture the solar corridor–enabled synergies and put it all together to deliver those results for various real-world production environments and provide the framework for success.

Based on our recognition and acceptance of the preceding spatial arrangements that embrace the "core principles" of the solar corridor crop system (SCCS) and the requisite understanding of the voluminous research findings in all disciplines that have led to the unprecedented crop yields represented in our "controls," we can begin the uniquely productive process of "putting it all together."

The objectives of this chapter are as follows: (1) to review the proven results and their complementary values as potential incentives to deploy the significant extra effort required to establish the SCCS; (2) to establish the spatial architecture that may require a significant sea change in perspective and, indeed, a 180 degree paradigm shift; (3) to recognize and acknowledge the importance of best-management practices (BMPs) that

The Solar Corridor Crop System
DOI: https://doi.org/10.1016/B978-0-12-814792-4.00005-X

advanced your sustainable productivity of corn to this point; and (4) to spell out the operational sequence for establishment, through deployment of your SCCS.

For you to maximize both the sustainable and immediate practical[1] potential of the crop production resources your land has access to, and to achieve the challenge acclaimed in *Gulliver's Travels*—not only at the level of the day but from today's justifiably acclaimed high level of productivity—we offer the SCCS for your consideration. The solar corridor puts you in a position to more fully access all of the natural resources your land has access to. Corn planted in the SCCS pattern is depicted in Fig. 5.1 adjacent to a corn monoculture. The solar corridor puts all of your chloroplasts in position to access any incident sunlight each day that the sun shines. The chloroplasts are somewhat like the engine in your tractor. They receive the atmospheric-sourced photochemical energy source as incident sunlight that "energizes" plant growth. Once "energized" the chloroplasts enable the biocapture of atmospheric-sourced CO_2, which is somewhat like the fuel your tractor engine needs. The resulting photosynthate sourced carbon compounds constitute the largest component of crop dry matter yield (see Fig. 2.7; Loomis and Williams, 1963).

Figure 5.1 Depiction of a 60-in. corn test adjacent to a corn monoculture. *Photo courtesy Bob Recker, cedarvalleyinnovation.com.*

[1] Feasibly and infinitely accessible.

Historical perspective

Since the beginning of time[2] incident sunlight, the premier photochemical energy source, and atmospheric carbon dioxide (CO_2), the largest building block (or consumed component) of crop yield, have been abundantly accessible and available for crop production. The remarkable increases in crop production so far have been achieved, largely, by the successful manipulation or satisfaction of the other crop production factors, thus leaving most effective plant access to incident sunlight and the subsequent biosynthesis of atmospheric CO_2 as the new most limiting factor in modern best-managed cornfields where all of the other limiting factors have been satisfied. Accordingly, the objective of the properly designed, developed, and deployed SCCS is to maximize the most effective plant access to (or reception of) incident sunlight and its contribution to photosynthetically active radiation (PAR) and radiation use efficiency (RUE) in its role of maximizing the subsequent biosynthesis of atmospheric CO_2 and transforming it into crop yield, enhanced soil health and sustainability of the farm, and integrated community.

As an interdisciplinary corn production agronomist, the author provides front-end insight and developmental experience in developing and capturing the solar corridor—enabled synergies. This experience was relied on to introduce a portion of the potential system options that may be deployed under a myriad of crop production conditions and challenges. These challenges were both of the real world and those artificially created by the continued pursuit of the 150-year-old paradigm shift established after the colonists broke through the Cumberland Gap and discovered more good arable farmland than they could farm. Subsequently, Cyrus McCormick, John Deere, J. I. Case, and an appropriately incentivized Farm Equipment Industry led a very effective response to eliminate the problem, even while perpetuating the paradigm. Analyses of the solar corridor recognize and respond to the reality that the ratio of good arable farmland to people potentially capable of farming it is now reversed, and the amount of good arable farmland is now recognized as the finite factor, that is, "They ain't making any more good farmland!"

It predicates, at the outset, that the preferred and/or ultimately adopted version of the solar corridor depends on the vision, immediate

[2] Genesis 1:3.

objectives, and specific production environment of each producer. Indeed, its adoption requires at least a 180 degree, if not a 360 degree, paradigm shift that may be too much to expect in this generation. Those who run the preliminary hybrid screening trials will have an informed knowledge of its potential performance in their operation. Those who don't, won't. In this chapter I will try to add to that informed knowledge base as it is applicable to some of the many specific production environments where the SCCS enables uniquely successful outcomes. Certainly as more readers become involved and share their findings, it has the potential to become easier to access the needed level of relevant information. I see how corridor options can perform under other production environments as relevant. Anything less is not relevant to the enormity of the challenge tackled by this chapter.

Explanation/details

1. The SCCS yield data speaks for itself. With recognition and acceptance of the hybrid-specific nature of the yield data, it demonstrates a potential sea change increase in spatial productivity and net operating profits (NOPs) from every acre or hectare where the SCCS is successfully applied.
2. *Spatial architecture.* Traditional state-of-the-art BMP crop rotations have one field, farm, or otherwise segregated block of land for each crop in the rotation. The 2-year rotation adapted to the SCCS (Fig. 2.8) instead shows both crops in the same field, for a complete rotation every 60 in. Thus there are no field boundaries between crops in the same rotation. This example leaves natural 52.5-in. wide (solar) corridors between each corn row that are wide enough for incident sunlight to reach every chloroplast on 8–9-ft.-tall corn plants. (Shorter plants don't need quite as much width and taller plants benefit from a little more width.)[3] The corn-compatible cultivar(s) growing on the "floor" of the solar corridor utilize the incident sunlight that falls between the corn rows. This significant conceptual change on how we visualize crop production in this crop system may be the greatest impediment to a full understanding of, and subsequent acceptance of, the nearly inconceivable additional values enabled by appropriate

[3] See US Patent no. 6052941 (Deichman, 2000).

deployment of the SCCS. The design shown in Fig. 2.8, however, is just one example of one of many solar corridor configurations and/or crop option versions of the SCCS. For any given row width within our extensive experience with 30–38-in. rows, maximum productivity is positively correlated to frequency of the SCs in the field. Fig. 2.8 illustrates the maximum yield frequency (every other row) and was the author's refinement of the protocol used for the seminal studies reported in this seminal book. For example, our corn yield results have consistently produced 2× corn yield without the precise diamond-spaced diagram of the corn that may be attained with precision planters that we didn't have for our studies, and projects a 1× winter wheat or other winter grain yield from three 10-in. rows and a 2× stubble crop yield on the other half of the field. With three winter grain rows, the center row fits nicely between the twin corn row immediately after corn harvest and occupies exactly half of the field. The stubble crop occupies the same space as the winter grain.

For ease of application, but with a much more limited indication of the yield results this book addresses, the opposite end of the frequency spectrum may be an alternative you may choose to begin with. For example, you may have one SC on each end (EE) of the planter swath, which could be accomplished with only one setting change to the planter marker and simultaneously reduce the planting (time) and harvesting (time) based on a reduction of swath numbers to plant the field. (A harvest-sized swath of soybeans, hay, turnips, pumpkins, etc. would be appropriate to flank each side of the EE example.) An improved indication of yield improvement on the above would be to add two SCs per planter swath flanking both sides of one of the two center rows of the planter that should indicate how the corridor on both sides of the row increases production versus one side of the row as in the EE example. While not a valid replacement for the preliminary screening trial (PST)—if that swath produces as much as a properly bordered (one planter pass or at least six rows on each side) control swath with every row in place—you may have found a hybrid that would be a potentially good hybrid for the SCCS.

In my first PSTs as described in Chapter 2, "Variety selection for the solar corridor crop system" (Deichman, 2019), a SC on only one side of the row was my ½T(e), ½T(w), ½T(s), or ½T(n) treatment, depending on which (direction) side of the row the SC was on. The overwhelming results of those PSTs indicated an SC on both sides of the row was required to maximize productivity of the SC and the ½T was not proposed as a treatment in our advanced seminal studies. Thus all of our advanced seminal study treatments had T1 (single row) or T2 (twin row) with a SC between every row. After several years of PST and subsequent field trials and 1 year in our advanced seminal studies, with results that strongly supported the twin row, the single row (T1) was discontinued in favor of the twin row treatments which maximized SC corn yield. Thus the treatments presented in all seminal papers presented by me, except as noted, were twin row treatments with a SC between each twin row. For comparison of twin to single rows, Row Treatment 3 at Nelson 2 is the single row or T1 treatment in the spreadsheet (www.agronomy.org/files/about-agronomy/data-from-1992-sccs-trials.xlsx) which you can review by each hybrid and each population as you choose. We're not denying that T1 (single row) with closer plant spacing within the row will work; we're reporting on the work for which we have valid data. (All of our solar corridor treatments versus controls were of the same population.) We split each of our solar corridor treatments by population treatments to overview the hybrid-specific SC by population responses within each hybrid that we evaluated. We encourage research on single rows that our seminal work didn't pursue. Two of the solar corridor papers presented at the 2018 American Society of Agronomy (ASA) meetings reported results with single rows. One featured closer in-row spacing (Recker, 2018) and the other was a PST or PST with the same in-row spacing, thus at one-half as many plants per acre (Reinbott and Kremer, 2018). Detailed information you will need for your subsequent preference decisions will follow.

We understand that everyone may not choose to, or be prepared to, deploy my max yield performance preferences all at once and recognize that a more gradual transition has many practical advantages. At this point, we just want you to be fully aware of the corn yield results of the basic agronomic performance parameters we measured, of the SCCS and its potential deployment under your production environment. We don't, however, want anyone to draw preliminary conclusions resulting from shortcuts taken in the process. Based on our recognition and acceptance of the preceding spatial arrangements that embrace the "core" principles of the SCCS and the requisite understanding of the voluminous research findings in all of the disciplines that have led to unprecedented crop yields represented in our state-of-the-art BMP "controls," we can begin the most productive process of "putting it all together." Due to time and space constraints of this seminal book on the SCCS, anything more than a cursory review of this voluminous research and the cumulative experience of the yield increases in one lifetime is beyond the scope of this volume. In the case of corn, yield increases of 130 bushels per acre has occurred in the United States in the last 80 years. In your area of the world it might be even more or an even greater proportional increase. To put all of the above into its practical application, we illustrate with a modern farmer well-versed in the art and science of corn production in the US Corn Belt and ask him/her to use every production practice that has proven to be effective in maximizing site-specific corn yields under his/her production environment and simply add the SC treatment in small strips where their SC treatment yields can be compared in his/her fields with those of your current state-of-the-art BMPs (control)s that have proven to be your most productive management practices. If those production practices have balanced your soil pH and soil biology with soil/plant nutrients including water and oxygen at levels that eliminate all of them as yield-limiting factors, you have properly set the table to apply the solar corridor to further increase your NOPs, corn yield, and carbon sequestration, etc. per corn acre, and potentially enhance the sustainability and health of your most finite crop production resource—your arable soil—by adding and successfully integrating the SC into your production practices.

Phase 1: spell out the operational sequence

To the extent that the above yield-limiting factors are satisfied, make an orderly transition to the SCCS as determined by the specifics of each individual situation. For descriptive purposes, we'll call this evaluation or testing process Phase I. Once completed, Phase I will provide the critical information needed to make an informed decision on its potential productivity on your farm—anything less does not.

First, deploy the PST as described in Chapter 2, Variety selection for the solar corridor crop system, to ensure that you are working with SC responsive hybrids. Though our PST described standards call for the 85% and 90% criteria, pending your personal identification with those hybrid (s), proceed with the following examples.

Real-world example 1

This real-world example may be best suited to maximize personal and enable mutual community benefits from the SCCS that we will address in Chapter 8, "Vision." If you are a small acreage family farmer, beginning a new family farm, or, simply put, a farmer searching for more actual NOPs who could handle "more acres" but do not have "more acres," the SCCS offers a unique 180 degree crop production paradigm shift that enables a sea change in spatial unit productivity, profitability, and sustainable opportunity.

Fig. 2.8 illustrates a version of one of those paradigm shifts that could work well on all cropland suitable for corn and winter grains or oilseed crops. On land where erosion potential is minimal the winter grain could be replaced with a large variety of high-value community-supported agriculture (CSA) crops that are ready for harvest before the corn begins to capture the bulk of the incident sunlight and later with fall crops (turnips, etc.) for produce, pasture, and/or cover crops. If the main crop is soybean, sweet corn is an excellent soybean-compatible CSA crop. If the main crop is corn, the sweet corn needs to be a very early variety. For many reasons, most of which are obvious to the astute reader, this chapter is written with the objective of identifying some of the multitude of options opened up by the SC concept rather than detailing the full pros and cons of each. I have no doubt that those who have studiously and courageously accepted this challenge are knowledgeable of the various crops suitable to their growing season and their chronological development sequences and are skilled in the art of crop production, can fill in the blanks to determine the crops that best suit the specific situation. If the wheat and amazingly robust stubble (RS) crop that the enriched solar corridor production environment enables causes insurmountable harvest impediments, so many details yet to be worked out, when ready, will be in the second edition of this book. When (or if) an improved version of the Landec polymer soybean seed coating becomes available you will not have to be concerned with the soybean planting date. Unfortunately we don't have a corn main crop/soybean floor crop system worked out. I see a potential route to it but don't yet have the funds to research it. Maize Research Unlimited uncovered the route to pursue it, but the route is still untraveled. Also, sorry to say (except in a few community-oriented communities) local CSA markets have not encouraged the farmer to produce for them. As it is for every situation, the PST needs to be conducted first. Pumpkins, turnips, oats, clover, etc. in your PST could give you a heads-up on your next step. If you want to do this, try to contact me first.

Real-world example 2

For the typical one-to-two-hectare farmer in a developing nation such as Malawi, the sky is the limit. This is in regard to the vital impact on human nutrition, variety of floor crops, and current corn yields without constraints imposed by capital costs invested in large-scale farm equipment. Further details can be found at Lilongwe University of Agriculture and Natural Resources (LUANAR) website: www.bunda.luanar.mw/luanar/. This is where the SCCS may potentially have the most direct impact on local food scarcity.

Real-world example 3

The large corn grower who has invested in large-scale equipment may have a significant additional hurdle. Currently available modern capital-intensive equipment was designed as the ultimate deployment of the existing 150-year-old paradigm (sourced from the revelation discovered by the colonists when they broke through the Cumberland Gap)—not the solar corridor paradigm. At that time the vast new expanse of land to the west opened up a new arable land to people (L:p) ratio, which disclosed more land than the number of people who were capable of farming it. Now the ratio is reversed; it is now P:l. Due in no small part to our large-scale equipment–driven 150-year-old crop production paradigm, that we are still pursuing, we have more than enough people capable of farming it. Arable land is now our most finite crop production resource. Thus in a world where (1) people are starving to death; (2) we still depend on fossil-sourced carbon for personal automotive fuels; and (3) our finite high-quality arable land or soils base is shrinking, the focus on more production per spatial unit via the SC paradigm shift offers a logical, if not currently popular, problem-solving approach.

Equipment for the solar corridor, not the subject of this first edition book, hopefully will be in the next edition. At the outset, I envisioned that with a proactive agricultural equipment industry and well-equipped farm shops, the equipment hurdle could be successfully and expeditiously navigated. Basic easily adapted options were based on five observations. (1) Nearly every farm is home to a motorized chassis (four-wheeler) with a footprint capability of navigating between the wide corn rows. (2) Nearly every farm has access to retrofittable obsolete-sized combine thresher units no wider than 40 in. (3) Commercial high-clearance chassis are routinely produced and used in modern corn production practice. (With a grain tank instead of a spray tank, wheat stripper heads instead of a spray boom, and a small threshing unit we have a nucleus to work from.) (4) Previously available flail forage choppers and direct-cut forage choppers are available. (5) Soybeans, with sufficient agronomic work, could serve as a main crop and accommodate any combine platform width.

Subsequently, three more potential options were added: (1) Thanks to Rick Nelson and Bob Lindgren the Nelson Farm Shop retrofitted a combine using obsolete components off the shelf and from the scrap heap. This combine successfully harvested main crop corn and floor crop soybeans. (2) A major land grant university ran a Capstone project that, pending commercial demand, indicated the equipment could be commercially produced. (3) Improved stripper heads for small grain.

Real-world example 4

The livestock producer can overcome the harvest equipment hurdle by using livestock to harvest the floor crops after corn harvest. Conversely, I know of no greater value for the winter small grain and SC enhanced RS crop than chopping it for ruminant livestock. This book is not prepared to further discuss livestock beyond some ideas that may merit investigation, for example, "I've heard that sheep do well-pasturing corn leaves off of standing corn." The solar corridor (still green) corn leaves have completed their job by physiological maturity, giving a small window before grain harvest for extra value. I've never researched my hypothesis that some variation of the wild turkey or domestic poultry breed could harvest the RS crop while eating corn insects and fertilizing next year's corn row without the proven chicken tractor (Padgham, 2006), with chickens.

Real-world example 5

The farmer on fragile erosion-prone land can implement the SCCS and use the corn main crop and cover crop floor crop as a win/win benefactor specifically for the soil, operator, and grazing livestock.

Real-world example 6

The Integrated Community that supports the CSA and value-added concepts. An integrated community as envisioned by this author is a community that adds appropriate values to its community-sourced productivity before it leaves the community. This expansive topic will be developed more in depth in Chapter 6, "Socio-economic impacts" (Hendrickson, 2019), and in Chapter 8, "Vision" (Deichman and Kremer, 2019).

Significant reasons for taking the extra effort include the following. We consistently:

1. produced as much or more corn from one 60-in. row as we produced from two 30-in. rows; or

2. increased spatial corn yield over currently accepted BMPs or current state of the art corn rows;
3. from low-cost abundant atmospheric resources enabled by the SC, therefore, enabling more NOP per corn acre; and
4. depending on the crop options utilized, we produced additional production values from the other half (or vacated rows) of the field without reducing corn yield).[4]
5. Enhanced soil health (SH), water and air quality,[5] and
6. increased potential for future generations to maintain an economically viable place on the farm is among those additional values.
7. Overwhelming but anecdotal observations demonstrate unprecedented resistance to lodging from wind and stalk rot.[6]

Highly significant equipment concerns

Where corn is the main crop, unless the "floor" or rotational crop is a "pick-your-own" or "you-pick" CSA (community-supported agriculture) or a cover crop that is harvested by pasturing livestock, additional equipment is needed to harvest the floor crop. Where corn and soybeans alternate rows as the main crop, the SCCS requires a small forage chopper or combines with a 7.5–9-ft. footprint to harvest specific winter small grain over the top of soybean. Where soybeans are the main crop, no additional harvest equipment is required for winter small grain harvest. Whether it is the advanced modern large farmer or the 2.5-ha farmer in the least developed parts of the world, the SCCS can deliver unprecedented production increases.

1. The small farmer in the least developed countries does not encounter the mechanically based capital investment impediment to deploy the SCC, as the large farmer encounters. Potentially unnecessary bureaucratic problems and the legitimate economic burden of moving the food to the starving is mitigated, thereby eliminating starvation by increasing production where the hungry live.

[4] See discussion of floor crop production without affecting corn yield in Crescent City trials, Chapter 2.

[5] See discussion of soil health, environmental quality, and ecosystem services impacts of SCCS, Chapters 4 and 8.

[6] See discussion in "Fact 6" presented in Chapter 2.

2. Where officially designated "food deserts" occur in the United States or where people go to food pantries for food, we have a specific SCCS adaptation that alleviates those local problems on a local basis. Where small rural towns are suffering from years of what appears to be endless decline with no significant hope of anything to end the decline, the SCCS enables a 360 degree paradigm shift that can reverse the decline and exodus from the family farm into a newly reenergized community (RC). We will discuss this further in Chapter 6, Socio-economic impacts, and Chapter 8, Vision.

After your PST-identified hybrids are chosen, a simple way to begin your first SC follows. At fall planting use a 10-in. grain drill and alternate every other three rows of wheat or other true winter grain with spring oats and sow them equally divided: half wheat and half oats. Plant corn in the center of the oats swath. Ideally the spring oats will winter-kill and leave a relatively weed-free ground cover mulch for corn planting, which emerges into its SC-created microclimate further characterized by the swaths of taller winter grain and stubble crops on both sides of the row. (No more typical corn leaf damage, which was previously thought to be not significant, by airborne soil particles.) Adequate description of this microclimate effect defies my capability, and it is perhaps best left to your personal observation and for others to measure the specifics. If a 10-in. (or row spacing of one-third the main row crop spacing) drill is not available, you should plant corn first and drill the center wheat row as close as you can to the center of the twin corn rows immediately after corn harvest and defer the appropriately spaced drill purchase, pending your final decision on deployment of the concept. Double-plant your corn, likewise deferring our twin-row planter and wheat combine or forage chopper acquisition, pending your final decision on deployment of the concept. In the meantime, observe and record the RS crop growth and winter wheat to ensure that grain yield was not diminished and mow as necessary to control weeds through corn harvest. As an option, a good poultry flock, while helping the main crop via pest control, manure distribution, etc., could flourish in this production environment, followed by larger livestock after corn harvest (or after physiological maturity for sheep).

Our version of the SCCS uses currently popular row-width equipment (×2) to make it work based on widths of 30-in., 36-in., etc. Thus if you currently have 30-in. corn rows, it would be logical for the SCCS to have 60-in. rows as depicted in Fig. 5.2. If you have 36-in. corn rows, it would be logical for your SCCS to have 72-in. rows. Again using

Figure 5.2 Aerial view of a 60-in. on-farm, corn strip test established in a field of monoculture corn. *Photo courtesy Bob Recker, cedarvalleyinnovation.com.*

Fig. 2.8 illustration as our example, immediately after corn harvest, no-till the winter grain crop with the center winter grain row centered in the 6- to 8-in. space between the twin corn rows with the stubble/cover crop species seeded within the 30-in. winter grain swath. Eighteen months later, in the spring, plant your corn into the center of the 30-in. cover crop swath.

Developing a solar corridor crop system that works on your farm

At the beginning we acknowledged the critical value of current state-of-the-art, best-management practices and chose not to be redundant by detailing them in this volume. We will, however, clarify by saying every practice you've found to be successful at balancing your soil at the highest levels of production on this site is what we choose to build on. That's the point where we like to begin to develop the solar corridor. Setting your goal quantifies your intent and serves as a specific basis for all of the prescribed supporting practices and their integrated interdisciplinary synergies. No one else needs to set another's personal goal, but if your goal includes higher yield, increased NOP per finite spatial land resource unit or leaving your soil in a more productive state, the SCCS offers an unprecedented opportunity. Accordingly, setting your goal is the way to begin your journey of "putting it all together" for you and those who will follow you on this land.

This first step is so critically important that, before proceeding, I want to reemphasize with the following summary. Do everything you currently know that produces high corn yields on your site and screen high-yielding varieties at their state-of-the-art population or 30,000 plants per acre, whichever is highest, in harvest size strips for "ear flex" with every other row not planted. (Another preliminary screening method that offers operational advantages is detailed[7] in one of the pending patent applications and can be offered upon request.) The variety or varieties that produce 85% as much corn as the highest-yielding variety of equal critical flowering and developmental through physiological maturity dates under the above production environment, from half as many plants, meets our flex requirements to be an approved hybrid for your SCCS. If it is the only variety with those critical morphological parameters, the requirement is 90% of its control yield, at good yield levels [from half as many plants (rows)] without changing spatial positioning distance in the row.

Maturity rating alone provides an insufficient basis to evaluate hybrids for their potential performance in the SCCS. To improve the reliability of our results we require that both flowering dates and black layer be the same during the growing season of the PST for the performance of one hybrid to be compared to another. We shouldn't expect anyone to develop their SCCS until they see it will work under the specific production environment on their farm. Accordingly, we propose the first year to be a trial year to determine the farm's potential productivity, which also offers a positive response to our recommendation that you find a corn variety (or varieties) that is responsive enough to produce as much or more corn on one wide row as it produces on two conventional rows. One of many ways to do this is to ask your seed supplier for a dozen or more test size samples of hybrids that demonstrate uniquely high-capacity reproductive sinks and, beginning with the first planter row on the left, plant every other row of one planter swath across your field into Little Marvel peas, and plant every other row into corn with each corn row being a different corn variety of similar flowering dates. When you reach the end of the row, turn around in the direction that would place the two rows of peas side by side, but before starting back, replace the peas

[7] Third parties have invested in this screening method which removes spatial variability and lessens the risk but requires more work.

with the same variety of corn that is on its right. Repeat with each row of peas until you have two rows of each hybrid. Then replace the peas with corn seed of the adjacent row and make a planter round on both sides of your corn/pea treatments until all varieties are planted in the "control" portion of your tests. The first six rows out of the treatments should be reserved as border rows and not be a part of the measurements. (If using a four-row planter, go ahead and accept that as the border swath and asterisk the yield of the first two rows.)

You may offer 80%–99% of the peas to your local food pantry, 4-H, Future Farmers of America (FFA), church or food charity, etc. who may offer it to those who come and harvest the peas. If the replies aren't sufficient to harvest all of the peas at the same time, plant several varieties of peas and snow peas to stretch out the harvest interval as necessary to get them all harvested at the right maturity. After pea harvest you have the option of seeding those rows to legumes, turnips, radishes, or other cover crops of your choice. Make sure that weeds aren't allowed to establish and take advantage of the extra sunlight; we want to use the extra sunlight to increase crop yield. Take corn population counts when most convenient. Each row that has two corn plants per foot of row should be of sufficient population for a good PST evaluation of that variety. To the extent that population is lower than optimum, your positive response will be reduced by treatment (sunlight) contamination, thus negating the true value of the light-enriched solar corridor.

Soon after full black layer or physiological maturity or an R6 corn reproductive stage (Abendroth et al., 2011) but before combine harvest, you have the option of your well-supervised team to harvest and measure a length of row representing 1000th of an acre from uniform field sites of the same planter row where both the treatment and the control row each have the same plant count.

Conclusions

The solar corridor reorders the spatial arrangement of plants in the cornfield to create a unique canopy architecture that enables the growing crop to make more efficient use of the premier photochemical energy source that drives photosynthesis—incident sunlight. Approximately 90%–95% of the total dry matter of plants consists of photosynthate-derived carbon compounds originating from photosynthesis (Loomis and Williams, 1963).

Photosynthesis enables the uniquely creative biocapture, or biosynthesis, of atmospheric-sourced CO_2 (the plant's carbon source)—the largest component of #2 corn yield and uniquely valuable component of sequestration and subsequent synergetic soil enhancements, uniquely deposited relatively uniformly throughout the soil rhizosphere. Both incident sunlight and atmospheric CO_2 are infinite, obviously sustainable, feasibly accessible, and low cost. Unfortunately, current state-of-the-art, best management practices, high-yielding cornfields deprive mature chloroplasts of access to incident sunlight during critical reproductive stages of growth. The solar corridor corrects that basic yield-limiting factor. Atmospheric-sourced CO_2 cycling in the current carbon cycle is the same CO_2 cited as the major contributor to global warming. Relative to other cited serious global warming gases, CO_2 is very persistent.

Don't forget: The SCCS is hybrid specific. Use our PST or one of your own that meets our PST criteria. The trials can be run in replicated strips in corn production fields on your farm (Fig. 5.3). Assuming you find the hybrid(s) that meets the SC PST criteria for you, under your conditions, and you (1) do everything you've learned about producing corn under your conditions, and (2) effectively integrate the solar corridor concept into your new SCCS, we expect that you'll achieve at least the results that we achieved.

Figure 5.3 Field-scale strip test of 60-in. corn in Iowa. *Photo courtesy Bob Recker, cedarvalleyinnovation.com.*

Definition of terms

Double plant Planting the main crop (e.g., corn) again, creating a twin row 6–8 in. from the row planted by the first planter swath, immediately after the first round with the planter

Floor crop (FC) A crop growing on the SC floor between the wide rows. (From Jan. through May winter grain and stubble crops receive solar radiation benefits as the interim main crop.) Details indicate that it should (1) have a shorter stature to prevent intercepting any incident sunlight ahead of the main crop, or (2) utilize its incident sunlight during the main crop's off season. Ideally, it would be in addition to the main crop, like specific legumes capturing and providing atmospheric N, hosting nutrient-enhancing mycorrhizae, predator insects, etc. and other crops that capture (and sequester) carbon, residual mobile nutrients (that would otherwise be lost as possible pollutants), and precipitation where it falls and reduces soil erosion, soil compaction, diseases, and wind damage to emerging plants, for example, winter grain shielding corn from leaf damage and from blowing soil particles, etc. Nearly any annual winter crop whose harvest date precedes main crop canopy impact on floor crop yield and meets the main crop compatibility specifications could be a good choice.

Main crop (MC) The crop that is the primary beneficiary crop in the SCCS (e.g., corn, sunflowers, soybeans, C-4 sorghums) in rows wide enough for incident sunlight to reach the lower leaves throughout the productive stages of growth.

Main crop compatible A floor crop that does not impact negatively on main crop yield. (Or in the case of corn, et al. "corn compatible.") We have not investigated the crops that theoretically could (e.g., peas, etc.) have a positive effect on main crop yield.

Solar corridor (SC) A corridor between crop rows wide enough (apart) to enable incident sunlight to reach the lowest leaves of the (main) crop in the wide rows. It is the functional canopy structure that powers the SCCS yields.

Solar corridor crop system (SCCS) The synergistic crop production practice briefed in this book. Or it is the integrated, interdisciplinary crop system that is designed to maximize productivity and/or beneficial utility of the SC. When presented as SCCS it means a properly designed, developed, and deployed solar corridor crop system.

References

Abendroth, L.J., Elmore, R.W., Boyer, M.J., Marlay, S.K., 2011. Corn Growth and Development. PMR 1009. Iowa State University Extension, Ames, IA.

Deichman, C.L., 2000. Plant arrangement for improving crop yields. U.S. Patent 6,052,941. Date issued: April 25, 2000.

Deichman, C.L., 2019. Variety selection for the solar corridor crop system (SCCS). In: Deichman, C.L., Kremer, R.J. (Eds.), The Solar Corridor Crop System. Elsevier Academic Press, San Diego, CA, Chapter 2.

Deichman, C.L., Kremer, R.J., 2019. Vision. In: Deichman, C.L., Kremer, R.J. (Eds.), The Solar Corridor Crop System. Elsevier Academic Press, San Diego, CA, Chapter 8.

Hendrickson, M.K., 2019. The economic and social conditions of the solar corridor cropping system. In: Deichman, C.L., Kremer, R.J. (Eds.), The Solar Corridor Crop System. Elsevier Academic Press, San Diego, CA, Chapter 6.

Loomis, R.S., Williams, W.A., 1963. Maximum crop productivity: an estimate. Crop Sci. 3, 67–72.

Padgham, J., 2006. Raising Poultry on Pasture: Ten Years of Success. American Pastured Poultry Producers Association, Hughesville, PA, p. 264.

Recker, R., 2018. Sunlight utilization of wide row corn (maize) as an enabler of sustainable agronomic practices. In: Agronomy Abstracts, ASA-CSSA Annual Meetings, Baltimore, MD.

Reinbott, T.M., Kremer, R.J., 2018. Screening corn hybrids for the solar corridor production system. In: Agronomy Abstracts, ASA-CSSA Annual Meetings, Baltimore, MD.

CHAPTER 6

The economic and social conditions of the solar corridor cropping system

Mary K. Hendrickson
Division of Applied Social Sciences, University of Missouri, Columbia, MO, United States

Contents

Introduction

The solar corridor cropping system (SCCS) is an innovative farm management tool that could provide a number of farm and community level benefits both in the American "Corn Belt" and in smallholder farming communities in many different areas of the world. In this chapter, I will primarily address the current conditions of midwestern cropping systems and potential opportunities at the farm and community level that could accrue from this system, although it is important to recognize that many

The Solar Corridor Crop System
DOI: https://doi.org/10.1016/B978-0-12-814792-4.00006-1

of the economic and social issues faced by American farmers are similar for smallholder farms joining into global commodity value chains. In the industrialized agriculture and food systems that have developed over the last 150 years in places like North America, we have delinked places of production from places of consumption, with far-ranging social, economic, and ecological consequences (Lyson, 2004). In addition to American farmers, smallholder farmers across the world participate in a system of food and agriculture with long social and ecological feedback loops that mask the consequences of the productivist orientation adopted in agricultural commodity systems (De Schutter, 2017). While providing an excessive number of calories for humans, the food and agriculture system refined and expanded in the American "Corn Belt" is structurally less resilient to shocks (Hendrickson, 2015). The life chances of farmers, food and farm workers, the communities in which they live, and the wider environment in which they are embedded are impacted in myriad ways by the organization of the dominant agriculture and food system, particularly as it relates to North American row crop production.

The SCCS is an alternative that has been developed within the heart of the corn–soybean monoculture of the midwest United States. Since 1978 crop diversity across the United States has declined significantly, especially in the heartland region (comprising most of the historic Corn Belt).[1] This region had the lowest diversity in seven of the eight United States Department of Agriculture (USDA) census years between 1978 and 2012 (Aguilar et al., 2015). The majority of the nation's 90 million acres of corn are grown in the heartland region, with the bulk of corn products used for livestock feed or ethanol production.[2] This kind of agriculture specialization across broad scales has created a monoculture cropping system that relies on only two rotations. Monocultures negatively impact the provision of ecosystem services and biodiversity by requiring higher production inputs and simplifying the ecosystem in ways that "may lead to ... loss of quantity and quality of [ecosystem] products and services"

[1] The heartland region as defined by the USDA encompasses all of the states of Ohio, Indiana, Illinois, and Iowa, most of Missouri, and portions of eastern Nebraska and South Dakota, southern Minnesota, and southwestern Kentucky (Aguilar et al., 2015).

[2] USDA overview of feedgrains: www.ers.usda.gov/topics/crops/corn-and-other-feed-grains/background/. University of Missouri's FAPRI estimated that over 70% of corn grown in the United States in 2016/2017 was used for feed or fuel (www.fapri.missouri.edu/publication/2018-august-baseline-update/).

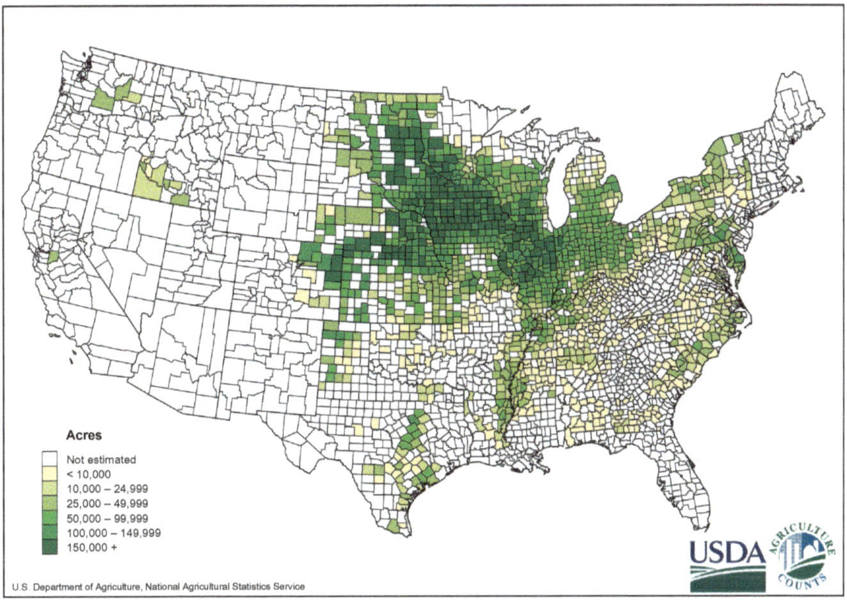

Figure 6.1 Corn for all purposes: planted acres by county for selected states, 2017.

(Klasen et al., 2016, p. 114). One need only glance at Fig. 6.1 to see the concentration of corn production across the heartland. Ecological and economic specialization go hand in hand in this case, with Klasen et al. (2016) raising concerns about the *scale* of specialization, particularly with reference to trade-offs between economic gains in relation to degradation of ecosystem services. Specialization on an individual farm might bring economic benefits without ecological harm, but across broad landscapes it may reduce ecosystem services. The key point is that the American heartland has specialized in row crop production that could threaten farmers and rural communities both economically and ecologically.

Specialization at such broad scales is difficult to undo. The question to be addressed here is could the SCCS provide economically and socially viable alternatives to corn and soybean monoculture for both US commodity farmers and smallholders around the world? Can this system enhance social, ecological, and economic resilience for farmers, and by extension, the communities in which they live? In this chapter, I describe the social and economic basis for current cropping systems in the Midwest, and their impacts on communities, before discussing how agronomic changes that produce the SCCS could potentially provide social and economic benefits, including community-level resilience.

Current conditions of US corn production

Monocultures are rooted not only in ecology but also in particular forms of social and economic organization that seek specialization and standardization. The kind of economic and ecological specialization apparent in the heartland's corn and soybean monocultures means that farmers face a series of consolidated markets in choosing what inputs to use on their farms and where to sell their products. In fact, some experts warn about "unprecedented consolidation" in several sectors, including seeds, agricultural chemicals, fertilizers, and animal genetics, accompanied (or driven) by increased concentration in the food processing and retail sectors (IPES-Food, 2017, p. 3). After the 2016–18 mergers of large agribusiness firms specializing in agricultural chemicals and seeds (Syngenta-ChemChina, Dow-DuPont, and Bayer-Monsanto), three entities may control as much as three-quarters of corn and soybeans planted in the United States (Begemann, 2017). In Table 6.1 the concentration ratio of the market shares of the top four and top eight firms in global seed and agrichemical scales are shown as they existed in 2011. In seeds, eight firms became six by 2018, while the top eight firms in agrichemicals became five (Table 6.1). Without accounting for divestures mandated by competition authorities in Europe and other places, the ratio of the global proprietary seed market controlled by the top four firms (CR4) increased from 54% to 61%, while in agrichemicals the CR4 increased from 62% to 82%. Consolidation in input markets is a global phenomenon; essentially corn and soybean farmers of the heartland face the same firms that farmers in other specialized grain regions like South America and Eastern Europe do. As firms merge, they tend to rationalize their operations by reducing redundancies in products, leading to fewer choices for things like seed, or plant protection chemicals. This results in limited choices for farmers of what to plant and how to manage that crop.

When farmers in the heartland sell their crops, they have limited choices as well. A few major grain traders like ADM, Bunge, Cargill, and Louis Dreyfus—and the emerging Chinese firm COFCO (see Belesky and Lawrence, 2018)—move the vast majority of grain that moves between nations, including the 10%–20% of the US corn crop that is exported. In the 1990s and early 2000s corn farmers across the Corn Belt participated in New Generation Cooperatives to process their corn into ethanol.[3] Since that period the ethanol industry has consolidated with the

[3] New Generation Cooperatives generally require capital contributions that provide and enforce delivery rights of the raw product that will be transformed into a value-added product. The farmer gets a guaranteed market and a share of the value-added profit. However, the cooperative one-member, one-vote custom still applies (Harris et al., 1996).

Table 6.1 Global seed and agrichemical companies.

Seed company and global rank on global seed sales	Concentration ratio of global proprietary seed market (2011)	Merger partner	Percent of global proprietary seed market of combined firm (estimated 2017)
Monsanto (1)		Bayer (7)	29
DuPont (2)		Dow (8)	18
Syngenta (3)			9
Vilmorin (4)	CR4—54%		5
WinField (5)			
KWS (6)			
Bayer CropScience (7)		Monsanto (1)	
Dow Agrosciences (8)	CR8—63%	DuPont (8)	
Agrichemical company and global rank	**Concentration ratio of global agrichemical market (current)**	**Merger partner**	**Percent of global agrichemical market of combined firm (estimated)**
Syngenta (1)		ChemChina (7)	29
Bayer CropScience (2)		Monsanto (5)	25
BASF (3)			12
Dow Agrosciences (4)	CR4—62%	DuPont (6)	16
Monsanto (5)			
DuPont (6)		Dow (4)	
ChemChina (7)		Syngenta (1)	
Nufarm (8)	CR8—87%		

Source: Hendrickson, M.K., Howard, P.H., Constance, D.H., 2019. In: Gibson, J.W., Alexander, S.E. (Eds.), Defense of Farmers. University of Nebraska Press, Lincoln, NE, based on market shares cited by ETC Group (2013).

top four firms (ADM, Valero Energy Corporation, Green Plains Renewable Energy, and Flint Hills Resources) controlling 40% of the market.[4] It is within these market conditions that today's American corn farmers must try to make a living, with attendant consequences on them, their communities, and their ecology.

[4] Information on market share comes from Farm Industry News in December 2016. www.farmindustrynews.com/ethanol/5-largest-ethanol-producers.

How did we get to this point?

In his excellent overview of changes in American agriculture, Lyson (2004) locates increased specialization and industrialization in the food system as emerging from three major agricultural revolutions in the 19th and 20th centuries. Increasing mechanization greatly diminished the need for human labor in the production of food. Most efficiency that has been pursued and perfected in the dominant agrifood system is the reduction of labor needed. Moving people out of agriculture changes the nature of relationships between production and consumption, including spatially lengthening these relationships. In other words communities no longer have to rely on their agriculture base for food products; instead geographic regions across the country (and the world) specialize in particular kinds of production (e.g., grains in the Midwest, fruit and vegetables in the South and West). While many consumers enjoy the wide range of foods available to them, these changes have made farmers and the food system dependent upon fossil fuels, while also changing the relationship of consumers to their own and others' ecologies.

While mechanization reduced the number of people needed in agriculture, the end of World War II ushered in the use of chemicals which reduced our need for complex systems of management. Commodity agriculture systems have used chemicals derived from or through petroleum products to replace cover crops and crop rotations in managing soil fertility, as well as using chemicals to manage insects, diseases, and weeds rather than through tillage, rotation, succession plantings, and other complex management practices. This has led to monoculture in many forms but the minimization of complexity and the rise of standardization have also allowed the development of industrial organizations (agribusinesses like ADM and Bayer) that dominate today's commodity cropping systems. With a reduction in complexity at the farm level, the management skills of farmers are less necessary, decreasing the returns (profit) to management. In fact, a common refrain among farm management specialists is "Dumb it down and scale it up." While management skills like implementing crop rotations, succession plantings or cover crops are owned and deployed by the farmer; buying chemical inputs from off the farm ensures the farmer must now have access to a ready and cheap source of capital. Farmers in these commodity production systems end up manipulating financial instruments (e.g., using hedging or derivatives) to eke out profits (Clapp and Isakson, 2018), further privileging financial capital and distancing them from local ecological and economic relationships.

Lyson described the final revolution as the introduction of herbicide-tolerant and insect-resistant crops in the late 1990s. This revolution sparked the convergence of seed and agrichemical firms, leading to the dominance of three or four mega input firms, like Corteva Agriscience (Dow-DuPont's agricultural inputs firm) or Bayer (intensified after its acquisition of Monsanto). All three of these agricultural revolutions have spawned specialization at farm and geographic scales; for instance, row crop farms in the Corn Belt are entirely reliant on corn and soybeans for their livelihoods while the livestock that eats their crops are concentrated in massive operations in specific locales, sometimes far away from grain production (e.g., large-scale poultry and livestock operations in the southeastern United States).

As one can see from this brief history, much of the 20th century in American agriculture—particularly that practiced in the American heartland—employed a productivist orientation, meaning the focus is on producing as much food and fiber for as little economic cost as possible (Lyson and Guptill, 2004). In this system, farming is delinked from specific locations and their social and cultural relationships, with farmers participating mainly as workers and rural places useful primarily as production sites. Essentially farmers, communities, and ecologies are viewed as cogs in a manufacturing system that can stretch around the globe. Lyson and Guptill (2004) represent *civic* agriculture as a system where farmers situated in a particular place make their livelihood through providing for the needs of their own communities or regions. They note that while "commodity agriculture is incorporated into an increasingly global and corporate-controlled food system, civic agriculture is linked to local or regional markets, often through direct sales to consumers" (p. 371). The SCCS brings new possibilities for building bridges between these two types of agricultural systems, as it focuses on a commodity crop—corn—but opens spaces for alternative ways of producing it.

How well do farmers fare in commodity agriculture systems?

One way to understand the impact these industrialized systems have had on farmers is to examine how well farmers do at making a full-time living from farming. Since 2014 farm incomes have fallen 50% (Gloy, 2018). In 2016 US net farm income, in inflation-adjusted dollars, was at its lowest level since 2002, which was a time of below-cost-of-production crop prices supplemented with disaster payments to farmers.[5] In 2018 while farm income is forecast to be roughly the same as in 2016, corn receipts are expected to decrease from 2017, with some agricultural economists predicting in spring 2018 that farmers would lose $54 per acre on corn production in that year (Gloy, 2018). Adding to income worries in 2018 has been the disruptions in grain markets due to wider changes in American trade policy. Many heartland grain farmers bought machinery or expanded during a period of high prices from 2010 to 2014, when corn once reached a price of $8 per bushel, and are not financially structured for the low corn prices of under $3.75 per bushel of today. But it's not just about row crop farmers. Operators whose primary occupation is farming, but who gross less than $350,000 in agricultural sales, had a median farm income of *less than $500* in 2016 (Table 6.2). About two-fifths of the approximately two million US farms listed farming as the primary occupation of the farm operator; these roughly 800,000 full-time farmers can further be subdivided into those who generate gross farm sales exceeding $350,000 (commercial farms) and those who generate less than that (intermediate farms). While off-farm income kept the median household income of these farm households above the US median household income, the volatility of farm income means that "full-time farmers face greater household income risk" than farm households where the operator holds an off-farm job or their spouse works off the farm (Key et al., 2017, summary).

While this discussion has centered on all US farmers, a good portion of corn production in the heartland happens on mid-sized farms, identified by USDA as those with $350,000–$999,999 in annual gross sales. These farms are too large for emerging direct markets but are too small to compete well in global commodity chains, and are characterized as the declining "Agriculture of the Middle" (Lyson et al., 2008). While just 6% of all US farms, they accounted for about one-fifth of all agricultural sales and farmland. Four-fifths of cash grains production, which includes corn, happens on farms grossing $350,000 or more on agricultural sales (Figs. 6.2 and 6.3). Ninety percent of all family farms grossed less than $350,000, and while they operated about half of US farmland, they accounted for only 23% of agricultural production (Hoppe, 2017). In the

[5] Highlights from the August 2018 Farm Income Forecast by USDA's Economic Research Service: www.ers.usda.gov/topics/farm-economy/farm-sector-income-finances/highlights-from-the-farm-income-forecast/.

Table 6.2 Principal farm operator household finances, by ERS Farm Typology, 2016.

Item	Residence farms	Intermediate farms	Commercial farms	All farms
Number of farms	1,227,552	616,525	183,192	2,027,269
Income, median dollars per household				
Farm income	−2008	428	143,968	−940
Off-farm income	83,400	55,000	39,000	66,468
Earned income	65,000	30,975	22,500	40,000
Unearned income	24,962	24,962	6500	24,962
Total household income	84,511	57,594	199,738	76,250

Source: USDA Economic Research Service. Accessed at www.ers.usda.gov/webdocs/DataFiles/48870/table02.xls?v=0.

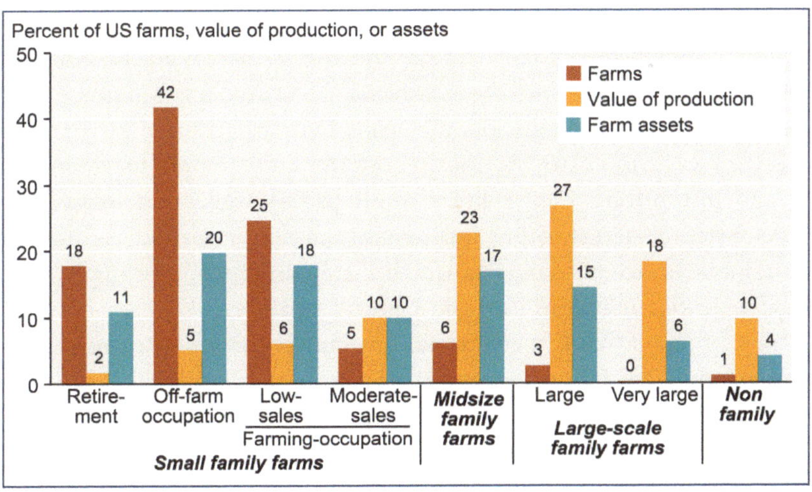

Figure 6.2 Distribution of farms, value of production, and farm assets, 2016. *Note:* Small family farms have annual gross cash farm income (GCFI) <$350,000, measured before deducting expenses. Midsize family farms have GCFI of $350,000–$999,999. Large-scale family farms have GCFI of $1,000,000 or more. *USDA, Economic Research Service and National Agricultural Statistics Service, 2016. Agricultural Resource Management Survey (data as of December 14, 2017).*

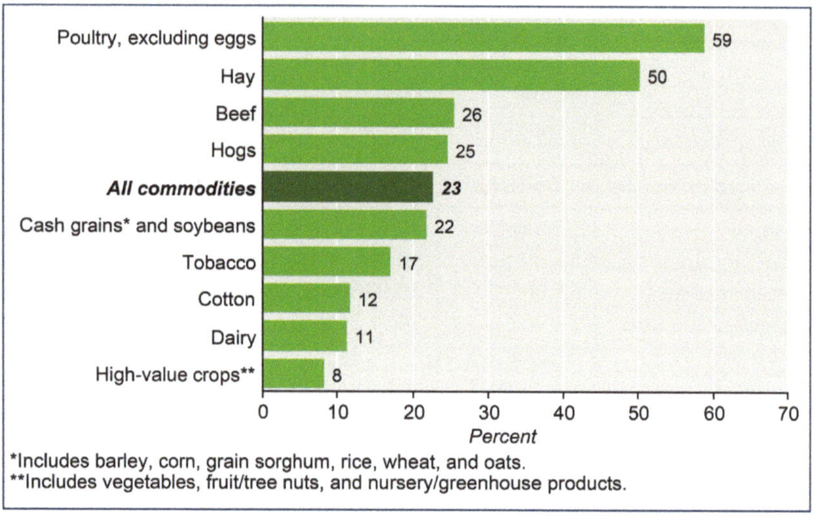

Figure 6.3 Share of the value of production from small family farms for selected commodities, 2016. *Note:* Small family farms have annual gross cash farm income (GCFI) <$350,000, measured before deducting expenses. *USDA, Economic Research Service and National Agricultural Statistics Service, 2016. Agricultural Resource Management Survey (data as of December 14, 2017).*

declining "Agriculture of the Middle" are the farmers who live and work in the rural communities of the heartland. Most mid-sized farms earn the majority of their household income from the farm (Burns, 2016), which means they contribute to the local economy and support local organizations and institutions. Lyson and Guptill (2004) argue that many of the linkages between farms, other businesses, and consumers are being eradicated for mid-sized and larger farms that are participating primarily in the *commodity agriculture* system (see also Legun and Bell, 2016).

One can argue that the economic structure in which row crop farming is embedded makes farmers vulnerable to more powerful actors in the system, such as mega input firms like Bayer or major processors/traders like ADM or Bunge. With my colleague (Hendrickson and James, 2005), I have argued that consolidated markets in the inputs required for industrialized farming and the market for agricultural products limit the decisions that farmers may be able to make about what crops and livestock they can produce and how they treat their land, animals, and workers. We also worry that constrained choices can force farmers into the "kinds of decisions that they otherwise would not have chosen for ethical or other reasons" (p. 283). For instance, grain farmers may be aware that using cover crops or adding a crop to the rotation would be beneficial for soil health but cannot obtain inputs or find markets for alternative crops because of market constraints, even when they may want to. Other researchers have reached similar conclusions. Stuart and Schewe (2016) documented that seed corn contracts in Michigan constrained the choices of farmers causing them to over-apply fertilizer to maximize yield, while Stuart (2009) found that farmers in the Central Coast of California felt pressured by their buyers to use practices they felt were ecologically destructive and unethical.

In a nutshell, for farmers, especially those operating mid-sized farms on which a great deal of corn is planted in the heartland, the current economic structure in which they find themselves embedded has had negative impacts on farm income, impacted the kinds of decisions they can make about what to produce and how to manage their ecosystems, and contributed to consolidated land ownership.

What happens to the broad landscape?

The ecology in which human society is embedded has also not been spared the negative impacts of the way we produce and consume food. Our current agricultural marketplace results in specialized monocultures and separates livestock from crop production both at the farm level and at larger regional geographies (Lyson, 2004). Livestock waste in these systems becomes an environmental pollutant rather than a recyclable resource contributing to soil health and fertility. Specialized markets mean that corn and soybeans become the de facto crop rotation across large portions of the Corn Belt, causing soil erosion in Iowa that costs farmers $1 billion per year (Eller, 2014) and contributing to the Dead Zone in the Gulf of Mexico. While cover crops, plant diversity, and reduced tillage can alleviate these problems, fewer farmers on the land farming larger acreages make the labor and timing of such practices challenging. Few if any markets exist for diversified crops and livestock meaning crop rotations are limited (Roesch-McNally et al., 2017), and input suppliers are oriented to providing seeds, chemicals, and equipment for the dominant crops in their region. Constrained choices for how farmers can manage their agroecosystems in relation to the structure of the markets they participate in leaves little room to maneuver ecologically in the face of the unpredictable temperatures, extreme precipitation events, or droughts.

Impacts on communities

The decisions these farmers must make also impact their communities. Scholars have long been interested in how the social and economic organization of food and farming impacts farmers and communities. Goldschmidt's (1978) seminal study of California agricultural organization in the 1940s found industrialized agriculture had adverse community effects while smaller-scale family-operated farms were associated with diverse economies and egalitarian social structures (see, e.g., Heffernan, 1984; Lobao, 1990; Lyson, 2004). In their meta-analysis on the relationship between agricultural structure and community well-being, Lobao and Stofferahn (2007) found detrimental effects of industrialized farming on communities were reported in 82% of 51 studies. These negative effects included greater income inequality or poverty, decreased retail trade and diversity of retail firms, population declines, and negative health effects of large livestock operations.

Of particular concern is the impact these effects have on different community capitals, or the relative stocks and flows of community-level resources that are available to mobilize for the common good (Flora and Flora, 2013). The Floras identify seven different capitals that communities can use (Table 6.3), including natural capital (soil, water), human capital (skills, abilities, and capacities), and social capital (the social bonds that let communities achieve goals). The community effects of the corn-soy rotation in the American heartland are often analyzed in financial capital

Table 6.3 Community capitals.

Financial	Money and other monetary instruments (stocks, bonds, shares) that can convert into other assets
Natural	Quality and quantity of water, soil health, biodiversity, landscapes
Social	Inclusive and diverse networks; reciprocity and trust; bonds between individuals and groups inside and outside community; collective action
Human	All the skills and abilities of people, leadership, knowledge
Political	Connections to people in power, access to resources, leverage, and influence to achieve goals
Built	Buildings and infrastructure such as roads, water and sewer systems, bridges, airports, utilities, hospitals, playgrounds
Cultural	Habits, heritage, generations, stories and traditions, spirituality, ethnicity

Source: Flora, C.B., Flora, J.L., 2013. Rural Communities: Legacy and Change, fourth ed. Westview Press, Boulder, CO.

terms—how much money do crop sales return to the community—but are less often considered in terms of human or social capital. For instance, what impact does the greater poverty identified in the meta-analysis by Lobao and Stofferahn (2007) have on the human capital of the American heartland, particularly in terms of skills and abilities? Or how does increasing income inequality affect the social bonds that let communities organize to achieve their goals (social capital)? These community capitals are bound up with the *civic agriculture* frame addressed earlier—when farmers are embedded in markets and social activities that are situated in a particular place they can contribute to the stocks of all the community capitals.

Broad scale ecological impacts

Another important community capital is natural capital, or the stock of healthy soil, clean water, breathable air, flourishing biodiversity, and attractive landscapes that a community has. The use of herbicides and pesticides in agriculture and mechanization has reduced the human capital necessary in large row crop operations, while also impacting natural capital. Soil loss is becoming a particular concern of farmers and conservationists. Fertilizer runoff has raised concerns about nitrate pollution in water systems across the Midwest (Vedachalam et al., 2018), even sparking a lawsuit against farmer drainage districts by the Des Moines Water Works (Eller, 2017). Weeds and pests have rapidly evolved resistance to widely used chemicals, with the introduction of herbicide-tolerant crops magnifying this problem in the last 20 years. In this case, natural capital is interconnected with social and human capital. Farmers who have replaced an integrated weed management strategy with chemical management decrease their own skill sets, leaving them floundering in the face of new resistant weeds. For instance, row crop farmers in the Southeast and Midwest have been scrambling in recent years to control a specific herbicide-resistant pigweed (*Amaranthus palmeri*), which thrives particularly well in a rapidly changing climate. In 2015 Monsanto introduced dicamba-tolerant soybeans and cotton as a way to control this and other resistant weeds; the problem is that dicamba can drift when applied under less than perfect application conditions, causing damage to a wide range of crops.[6] In 2016 and 2017 this damage caused considerable tension in rural communities, often pitting neighbor against neighbor; conflict over

[6] See blog posts from the University of Missouri Integrated Pest Management Program at https://ipm.missouri.edu/IPCM/2017/7/Ag_Industry_Do_we_have_a_problem_yet/.

dicamba damage was cited in the murder of an Arkansas farmer by a neighboring Missouri farmhand.[7] One can only imagine the economic and social pressures that manifested in that confrontation or the ethical conflict over farming choices that may have led to the situation.

Resilience in modern systems

Such ecological impacts can change risk and resilience equations for farmers, communities, and consumers. Resilience is generally defined as the ability of ecosystems and communities to decrease their vulnerability to sudden shocks, and to generate—through relationships and processes—options for adaptive change. In other words, how farmers and communities can deploy all their community capitals—natural, social, and human—matters for how resilient they can be in adapting to change. The industrialization and consolidation of agriculture has increased vulnerability for many farmers and communities and reduced the range of options for adaptation from field to plate (Hendrickson, 2015). With its lengthening supply chain, the industrialized agrifood system lacks tight ecological and social feedback loops, compromising adaptive responses that promote resilience (Lamine, 2015). Research in another industrializing system provides insight for what is happening in the American heartland. The Australian dairy sector deregulated to orient itself toward competing in the global marketplace. The changes increased competition between producers, which combined with the increasing power of supermarkets to dictate terms, led "producers to manage a system that carries increased risk in the physical environment and to animal health and welfare because of higher stocking rates and production per cow" (Sinclair et al., 2014, p. 379).

In this section I have painted a picture of industrialized agriculture in the American heartland that has had negative social, ecological, and economic consequences. How we have organized our agrifood system has forced many farmers out of farming as a full-time profession in the United States and decimated the number of smallholder farmers in other countries, while those who remain face consolidated markets that narrow the range of choices they can make about what kind of farmer they want to be and how they want to manage land, people, and animals. By limiting farmers' ability to manage their farms as an ecosystem, this consolidated and industrialized agriculture system has increased their vulnerability and reduced their resilience in the face of emerging threats, such as our changing climate. At the community level, the structure of our agrifood system

[7] See www.npr.org/2017/06/14/532879755/a-pesticide-a-pigweed-and-a-farmers-murder.

has reduced the community capitals that farmers and communities can mobilize to achieve their own goals and aspirations.

The possible farm and community level benefits of the solar corridor cropping system

> *Because the emphasis on corn and soybeans started in the twenties. It's not a development just of the last few years. Yes, it's gone crazy since the seventies, but we spent a century going down this road and everything . . . has been built around a certain way of doing things. On the one hand, if we take a look at it and say, 'This is not the way we should be doing things,' then, sure, we need to do things different. But on the other hand, I also don't think we should say, 'We should all start raising organics and we'll all just put on our overalls and we'll go out there and do it ourselves.'*
>
> ***Meghan Galloway, Nebraska farmer, quoted in Genoways (2017).***

As many readers understand, the current cropping system employed in the heartland region is supported by large financial investments in land, machinery, storage, and transportation by farmers, communities, and agribusinesses. Essentially we have invested all our community capitals—money, soil and water, and human knowledge and skill—in one way of doing things. Meghan Galloway, a sixth-generation diversified crop and livestock farmer in Nebraska, is making the point that reorienting such large investments cannot happen speedily or hastily. However, adopting the SCCS could provide a bridge for deploying community capitals in new ways, by supporting farmers in a gradual move away from the corn–soybean monoculture into new alternative ways of farming—even if still producing corn or other commodity crops. Using the SCCS, farmers could still engage in corn production for established markets while using the corridors for production that could accomplish other farm and community goals. Thus the SCCS could potentially provide social and economic benefits to farmers and communities, in addition to the ecological benefits addressed throughout the book.

Changing tastes, changing markets

Many consumers who are interested in alternative food systems often cannot understand why farmers don't change their farming systems to serve emerging alternative markets shaped by the demands of consumers for organic, natural, sustainable, or humanely raised products. After all, these markets are growing rapidly, especially among younger consumers,[8] with

[8] See Acosta's "Back to Our Roots" report available at www.acosta.com/news/new-acosta-research-dissects-the-profiles-of-today-s-natural-and-organic-shoppers.

over two-fifths of American consumers in a 2016 survey saying that sustainability had some to great impact on their food and beverage buying decisions (International Food Information Council Foundation, 2016). Young people are interested in changing the food system, and they often focus on transforming the large-scale monocultures of the American heartland or the large-scale livestock production such monoculture supports.

USDA estimated the market for local food—sold either directly from farm to consumer or through local food intermediaries—as $8.7 billion in 2015 (USDA, 2016). Local and regional food markets can have beneficial impacts on community capitals. In a survey of local food producers in Missouri and Nebraska, Rossi et al. (2017) found that farmers who were oriented to local food production and marketing produced larger economic impacts in terms of sales, GDP, and employment when compared with conventional agricultural systems (see also Jablonski et al., 2017). Schmit et al. (2017) noted increases in intellectual and human capital stocks in rural areas from farmers participating in alternative food systems in urban areas. However, much of the documented local food market activity is located near metropolitan centers, which disadvantages markets in more rural locales (Hendrickson and Massengale, 2017).

The amount of production that comes from the large fields and machinery of row crop farmers in the heartland makes it difficult for them to currently think about serving local food markets. However, by using the SCCS farmers will have strips of land in the corridors that provide space for experimenting with different food crops. Because farmers will not lose yields overall, they can explore different crops to help them access local or regional markets. In this way they keep 1 ft. in commodity production in order to maintain cash flow and other economic opportunities, while developing new income streams that allow them to access new markets. Because of the large scale of their existing crop farms, these farmers will likely need to focus on wholesale or institutional markets for less labor-intensive, more extensive crops. For instance pumpkins, sweet corn, some legumes or sweet potatoes can be done on a larger scale with less labor required for planting and harvesting as well as for pest and disease management. While the SCCS can open up some opportunities, other barriers to accessing these markets remain such as access to seed and inputs for alternative crops, lack of appropriate machinery or equipment, and farmer production knowledge for alternative crops.

The point here is not to change row crop farmers into vegetable farmers. By using the SCCS, there exists the potential for additional farm income from managing the corridors with alternative crops. The SCCS continues to provide crop income for farmers while they experiment with accessing different markets, which provides stability to the economic enterprise as well. On the other hand, farmers have to allow for the increased management costs of adapting new crops and enterprises on their farms. In this sense, stabilizing money flows can allow farmers to focus on increasing their management skills, reflecting increases in human capital. In addition, natural capital can be enhanced through the ecological benefits that accrue from SCCS (e.g., soil health). These natural capital benefits also yield economic value for the overall cropping system in reduced input costs for fertility treatments or managing crop diseases and pests.

Addressing food security

While many farmers may choose to use the SCCS for the ecological benefits it can provide, or for the kinds of additional income and new markets they may be able to access, some farmers may also consider how the corridors could improve food security in their communities. A paradox is that the places where a great deal of calories are produced also tends to have barriers to food access and availability. Rural Americans are more likely to be food insecure than suburban Americans, with only principal cities within metropolitan areas having higher rates than rural areas (Piontak and Schulman, 2014; Tanaka et al., 2014). Food insecurity is tightly linked with poverty; the nonmetro poverty rate was 16.9% in 2016, 3% points higher than the metro rate of 13.6%.[9] In addition, the consolidation of grocery stores that has taken place since the early 1990s has decreased the number of smaller stores that might serve rural areas (Piontak and Schulman, 2014). Food is often not available within these communities, and many families wouldn't be able to afford it if it were. Many rural communities in the heartland also lack emergency food resources such as food pantries that can help families in the region stretch their food supply. For instance, in northeast Missouri, the heart of corn production, some communities may only have pantries open once or twice a month for a very narrow window of specific times and days. These pantries also have little ability to accomodate the fresh foods necessary for health, a key component of human capital.

[9] For more information, see USDA Economic Research Service at www.ers.usda.gov/topics/rural-economy-population/rural-poverty-well-being/.

The SCCS could potentially open up acreage for production of vegetable (or possibly even some fruit) crops that could be used to address food insecurity within rural communities. For instance, food banks (which act as a wholesale distributor to retail sites such as food pantries or meal program) in South Carolina have been working with local farmers to move locally produced foods to their client agencies while also helping these farmers access other local or regional markets through transportation infrastructure (Robinson et al., 2013). Farmers could potentially use their corridors to produce high-quality staples such as beans or sweet potatoes that could be funneled through emergency food outlets. Gleaning groups could potentially work with some farmers to harvest crops grown in these corridors, reducing the need for extra labor on the part of the farmer.

Increasing community resilience through building community capitals

The resiliency of our food and farming system should be a key concern of everyone who eats. The processes that have created globally similar food production and consumption systems lack resistance to disruption, the latitude to accommodate change, and diversity of organizational forms, all key principles of resiliency (Table 6.4). It is a precarious system with few fail-safe mechanisms (redundancy of form and function). In terms of ecology, the consolidation of agricultural commodity markets, particularly for row crops, and accompanying concentration in agricultural input markets has led to decreased diversity of seeds and breeds; constrained biodiversity both above- and below-ground due to a reduction in complex management systems; and decreased the ability of farmers to make decisions relevant to their own ecological situation. In social and economic terms, consolidation in the agrifood sector has squeezed out smaller scales of farm and food businesses, and has led to the development of one organizational form—for-profit corporations—as the preferred vehicle for food production and distribution.

The challenge for creating resilient agrifood systems is that some actors, like international agribusiness firms, have more power to make decisions and choices about the food production and consumption than others, like farmers, workers, and consumers. The SCCS challenges this power in ways that create a more socially and ecologically resilient food system. For instance, the SCCS provides a bridge for farmers to stay in consolidated commodity row crop production, while making ecological changes that can help them adapt to changing climates and markets. Such a tactic also allows them to challenge the specialized cropping systems

Table 6.4 Dimensions of resilience for the agrifood system.

Dimension	Capacity	Current vulnerability
Resistance to disruption	What is the capacity of a system to fend off disruption?	Agrifood production and processing overuse of soil and water; greenhouse gas contribution to climate change
Latitude to accommodate change	What is the capacity of the system to stretch and accommodate change without it being disruptive?	Decreasing genetic resources for major crops and livestock; single-use production facilities; global production networks
Redundancy of role and function	Where are the overlapping systems that can back up each other?	Farm and geographic specialization; centralized storage and processing facilities; tight coordination and connections means few fail-safe mechanisms
Diversity of organizational forms	What is the diversity of scale, form, and organization across the food system?	Focus on efficiency, specialization, and standardization eliminates smaller and diverse forms of organization
Precariousness	How vulnerable is a system to losing its core identity and transitioning to a new state?	Food insecurity and hunger globally and within rich countries; industrial diet uses a great many resources while failing to provide adequate and healthy food for the majority

Source: Hendrickson, M.K., 2015. Resilience in a concentrated and consolidated food system. J. Environ. Stud. Sci. 5 (3), 418–431.

forced through consolidated markets (MacDonald et al., 2018) by developing new markets that create redundancy in the food system. By using the SCCS, farmers can increase their resistance to disruption by nurturing their soil and water resources; by using different genetic resources and encouraging biodiversity in the soil they also enhance their ability to accommodate change. The SCCS relies on intensive management strategies that could reduce off-farm, expensive inputs. If farmers are rewarded for those management strategies, profitability could rise and encourage stability in the face of the relentless march to larger-scale row crop production. By encouraging more farms, managed in unique ways, the SCCS can enhance the diversity of organizational forms in rural areas. In essence these steps toward resilience highlight how farmers and communities can mobilize human, natural, and social capital to make their farms and communities more resilient in the face of ongoing disruptions such as market consolidation, trade disputes, or climate change.

International approaches

Despite my focus on the social and economic conditions of cropping systems in the American heartland, many of the benefits of the SCCS are potentially greater for agricultural smallholders around the globe. Large-scale row crop agriculture has developed worldwide, with similar economic, ecological, and social costs as described above (see also Lowder et al., 2016; De Schutter, 2017). In fact, smallholder livelihoods globally are likely impacted even more negatively than in the United States, as these farmers live in areas where the consequences of climate change are more present, in societies with inadequate social safety nets. Applying the SCCS may provide an alternative for these farmers and their communities to embedding themselves in global commodity systems. For instance, smallholders may be able to better manage soil fertility on farm by rotating crops or planting cover crops within the corridors, rather than searching out often unavailable, always expensive synthetic fertilizer. This reduces their reliance on financial capital while allowing them to build human and natural capital (de Schutter, 2010). de Schutter (2010) also argues that reducing smallholder's reliance on global commodity chains will allow them to better serve local and regional markets, which can address many of the conditions of food insecurity which abound in their communities. Such linkages embed these smallholders in *civic* rather than *productivist* agriculture, building social capital and contributing to food system resiliency.

Conclusion

In this chapter I have argued that the *productivist* orientation in agriculture and food developed over the last century in American agriculture has had significant negative impacts on farmers, their communities, and their ecologies. Large-scale monoculture has developed in the American heartland, as well as globally, that has led to the dominant agrifood entities that constrain the choices of inputs that farmers can use as well as where they can market their products. These changes have disconnected farmers from local markets and consumers from their impact on ecologies, in the process negatively impacting human, natural, and social capital, the very resources that communities need for continuing resiliency.

What the SCCS provides is opportunity. Opportunity for American farmers to reorient their investments away from monoculture systems in ways that help build natural, human, and social capital while reembedding them in their local ecologies and communities. By providing a bridge between productivist and civic agriculture, the SCCS allows farmers to engage in new management practices that build their skills while protecting their natural capital. Perhaps the larger opportunity is for farmers to engage in new markets that can better serve their own families and communities, or to allow farmers to address issues of food insecurity in their own community. In international settings the SCCS can help smallholders develop viable livelihood strategies on their farms, rather than embedding them in global commodity chains in which they face uncertain futures. In these ways the SCCS provides opportunities to increase the resiliency of food and farming systems, without wreaking massive changes overnight.

Acknowledgment

This material is based upon work that is supported by the National Institute of Food and Agriculture, US Department of Agriculture, Hatch under 1002034.

References

Aguilar, J., Greta, G.G., Hendrickson, J.R., Archer, D.W., Forcella, F., Liebig, M.A., 2015. Crop species diversity changes in the United States: 1978–2012. PLoS One 10 (8), e0136580.

Begemann, S., 2017. What to watch as mergers near the finish line. *Agweb*. Retrieved August 15, 2018, <https://www.agweb.com/article/what-to-watch-as-mergers-near-the-finish-line-naa-sonja-begemann/>.

Belesky, P., Lawrence, G., 2018. Chinese state capitalism and neomercantilism in the contemporary food regime: contradictions, continuity and change. J. Peasant Stud. 1–23.

Burns, C., 2016. The number of midsize farms declined from 1992 to 2012, but their household finances remain strong. *Amber Waves*, December.

Clapp, J., Isakson, S.R., 2018. Risky returns: the implications of financialization in the food system. Dev. Change 49 (2), 437–460.

de Schutter, O., 2010. Agribusiness and the right to food. Report issued by the United Nations Special Rapporteur on the Right to Food. Accessed October 10, 2018 at <http://www.srfood.org/images/stories/pdf/officialreports/20100305_a-hrc-13-33_agribusiness_en.pdf>.

De Schutter, O., 2017. The political economy of food systems reform. Eur. Rev. Agric. Econ. 44 (4), 705–731.

Eller, D., 2014. Erosion estimated to cost iowa $1 billion in yield. *Des Moines Register*, May 3.

Eller, D., 2017. With water works' lawsuit dismissed, water quality is the legislature's problem. *Des Moines Register*, March 17.

Flora, C.B., Flora, J.L., 2013. Rural Communities: Legacy and Change, fourth ed. Westview Press, Boulder, CO.

Genoways, T., 2017. This Blessed Earth: A Year in the Life of an American Family Farm. W.W. Norton and Company, New York, NY.

Gloy, B., 2018. USDA corn and soybean 2018 cost of production: hope for profits? *Agricultural Economics Insights*. Retrieved August 18, 2018 <https://ageconomists.com/2018/03/19/usda-corn-soybean-cost-production-hope-profits-2018/>.

Goldschmidt, W., 1978. As You Sow:Three Studies in the Social Consequences of Agribusiness. Allenheld, Osmun and Co. Publishers Inc, Montclair, NJ.

Harris, A., Stefanson, B., Fulton, M., 1996. New generation cooperatives and cooperative theory. J. Coop. 11, 15–28.

Heffernan, W.D., 1984. Constraints in the U.S. poultry industry. Res. Rural Sociol. Dev. 1, 237–260.

Hendrickson, M.K., 2015. Resilience in a concentrated and consolidated food system. J. Environ. Stud. Sci. 5 (3), 418–431.

Hendrickson, M.K., Harvey, S.J., 2005. The ethics of constrained choice: how the industrialization of agriculture impacts farming and farmer behavior. J. Agric. Environ. Ethics 18 (3), 269–291.

Hendrickson, M.K., Massengale, S.H., 2017. Local food systems in North America: findings from Missouri and Nebraska. Proceedings of the XXVII European Society for Rural Sociology Congress. Institute of Sociology at Jagiellonian University, Krakow, Poland, pp. 222–223.

Hendrickson, M.K., Howard, P.H., Constance, D.H., 2019. In: Gibson, J.W., Alexander, S.E. (Eds.), Defense of Farmers. University of Nebraska Press, Lincoln NE.

Hoppe, R.A., 2017. America's diverse family farms: 2017 edition. Economic Information Bulletin 185. USDA Economic Research Service, Washington, D.C.

International Food Information Council Foundation, 2016. Food Decision 2016. Washington, D.C.

IPES-Food, 2017. Too big to feed: exploring the impacts of mega-mergers. Consolidation and Concentration of Power in the Agri-Food Sector. IPES-Food, <www.ipes-food.org>.

Jablonski, B., Hendrickson, M., Vogel, S., Schmitt, T., 2017. Local and regional food systems driving rural economic development. Harvesting Opportunity: The Power of Regional Food System Investments to Transform Communities. Federal Reserve and USDA, Washington, DC.

Key, N., Prager, D., Christopher, B., 2017. Farm household income volatility: an analysis using panel data from a national survey. Economic Research Report #226, USDA Economic Research Service. Washington, D.C.

Klasen, S., Meyer, K.M., Dislich, C., Euler, M., Faust, H., Gatton, M., et al., 2016. Economic and ecological trade-offs of agricultural specialization at different spatial scales. Ecol. Econ. 122, 111–120.

Lamine, C., 2015. Sustainability and resilience in agrifood systems: reconnecting agriculture, food and the environment. Sociol. Ruralis 55 (1), 41–61.

Legun, K., Bell, M.M., 2016. The second middle: conducers and the agrifood economy. J. Rural Stud. 48, 104–114.

Lobao, L., 1990. Locality and Inequality: Farm and Industry Structure and Socioeconomic Conditions. State University of New York Press, Albany, NY.

Lobao, L., Stofferahn, C.W., 2007. The community effects of industrialized farming: social science research and challenges to corporate farming laws. Agric. Hum. Val. 25 (2), 219–240.

Lowder, S.K., Skoet, J., Raney, T., 2016. The number, size, and distribution of farms, smallholder farms, and family farms worldwide. World Dev. 87, 16–29.

Lyson, T.A., 2004. Civic Agriculture. Tufts University Press, Lebanon, NH.

Lyson, T.A., Guptill, A., 2004. Commodity agriculture, civic agriculture and the future of U.S. Farming. Rural Sociol. 69 (3), 370–385.
Lyson, T.A., George, W.S., Welsh, R., 2008. Food and the Mid-Level Farm. MIT Press, Cambridge, MA.
MacDonald, J.M., Robert, A.H., Newton, D., 2018. Three decades of consolidation in U.S. agriculture. Economic Information Bulletin EIB-189. USDA Economic Research Service, Washington, DC.
Piontak, J.R., Schulman, M.D., 2014. Food insecurity in rural America. Contexts 13 (3), 75–77.
Robinson, K.L., Robinson, K.K., Carpio, C., Hughes, D., 2013. Linking sustainable agriculture and community development: the lowcountry food bank's use of locally grown foods. In: Beaulieu, L.J., Jordan, J.L. (Eds.), Developing Sustainable Agriculture and Community. Routledge, Oxford, UK, pp. 77–90.
Roesch-McNally, G.E., et al., 2017. The trouble with cover crops: farmers' experiences with overcoming barriers to adoption. Renew. Agric. Food Syst. 33, 322–333.
Rossi, J.D., Thomas, G.J., Hendrickson, M., 2017. The economic impacts of local and conventional food sales. J. Agric. Appl. Econ. 49 (4), 555–570.
Schmit, T.M., Becca, B.R.J., Minner, J., Kay, D., Christensen, L., 2017. Rural wealth creation of intellectual capital from urban local food system initiatives: developing indicators to assess change. Commun. Dev. 48 (5), 639–656.
Sinclair, K., Curtis, A., Mendham, E., Mitchell, M., 2014. Can resilience thinking provide useful insights for those examining efforts to transform contemporary agriculture? Agric. Hum. Val. 31 (3), 371–384.
Stuart, D., 2009. Constrained choice and ethical dilemmas in land management: environmental quality and food safety in California agriculture. J. Agric. Environ. Ethics 22 (1), 53–71.
Stuart, D., Schewe, R.L., 2016. Constrained choice and climate change mitigation in US agriculture: structural barriers to a climate change ethic. J. Agric. Environ. Ethics 29 (3), 369–385.
Tanaka, K., Patrick, H.M., Wolff, B., 2014. Food insecurity and obesity in rural America: paradoxes of the modern agrifood system. In: Bailey, C., Jensen, L., Ransom, E. (Eds.), Rural America in a Globalizing World. West Virginia University Press, Morgantown, WV, pp. 642–660.
USDA, 2016. Direct Farm Sales of Food: Results From the 2015 Local Food Marketing Practices Survey. National Agricultural Statistics Service, Washington, DC.
Vedachalam, S., Ankita, J.M., Heath, E.A., 2018. Source Water Quality and the Cost of Nitrate Treatment in the Mississippi River Basin. Northeast-Midwest Institute.

CHAPTER 7

The solar corridor crop system: a pivotal production system for African smallholder farmers

Kerry M. Clark
University of Missouri, Columbia, MO, United States

Contents

Introduction

Throughout the majority of human agricultural history, the production of more than one crop in a field was commonplace and necessary for sustained productivity. Legumes could help provide nutrients for grain crops and intercrops made better use of soil resources, held more biological diversity that helped keep disease and insects in check, and carried less risk for farmers when environmental conditions were unfavorable for one type of crop growth. But in many parts of the world, monoculture cropping has become the standard form of agricultural production over the past century. In a spatial monocrop, large land areas are planted with the same crop species, which allows for easy planting, harvesting, and chemical pesticide use with modern agricultural equipment. In agricultural systems where labor is more prevalent than mechanization, intercropping, the planting of two or more crop species in a land area, is more commonly utilized. In the United States, temporal crop rotation is usually

The Solar Corridor Crop System
DOI: https://doi.org/10.1016/B978-0-12-814792-4.00007-3

practiced with spatial monocrops, but with maize (*Zea mays*) and soybean (*Glycine max*) as the two major commodity crops, the rotation is limited and has led to ongoing problems such as decreased yields and increased success of soil-borne plant pathogens (Cook and Weller, 2004). In the solar corridor crop system (SCCS), crop yields are expected to improve due to changes in crop planting architecture that allows greater penetration of light into the crop canopy, leading to improved plant growth through increased photosynthesis (Kremer and Deichman, 2014).

In an SCCS using maize and soybean, planting configuration strategies can include the use of paired single or double narrow rows of maize intercropped between strips of soybean (Kremer and Deichman, 2014), or maize planted in ultra-wide rows of 152 cm (Nelson, 2014). In this regard, the SCCS is very similar to intercropping systems used by many smallholder farmers in developing nations around the world. Intercropping has a 1000-year-old history in China and is estimated to still be practiced on 20%–25% of arable land in that country (Knörzer et al., 2009). Intercropping is the dominant production system in many parts of sub-Saharan Africa (SSA), where up to 80% of nonirrigated land is intercropped (Jalloh, 2001). Approximately 83% of cropped land in northern Nigeria and 94% in Malawi is intercropped to a diverse group of plant species (Knörzer et al., 2009).

In economically developed nations, mechanization was an important factor in the industrialization of agriculture and switch to monocultures. The availability of credit and capital, favorable government policies, and access to inorganic fertilizer sources and agricultural chemicals also contributed to changes in farm size and practices (Christensen et al., 1964). In the United States the midpoint farm size has risen from 238 ha (589 acres) to 447 ha (1105 acres) in 1987–2007 (MacDonald, 2013). This is evidence of steadily increasing consolidation of land into larger farms, made possible by changes in technology and changes in farm organization that may limit financial risks (MacDonald, 2013). Many of these technological changes, such as larger and faster equipment, may hinder a move toward intercropping or an SCCS in the United States while some other technologies, such as geographical position systems and other information and monitoring technologies, may assist in making alternative production systems possible. The use of intercropping or the SCCS in technology-based agricultural systems will require new technological concepts to be explored or will necessitate changes to equipment or practices that were adopted for monoculture systems. The objective of this chapter is to demonstrate how the SCCS can be adopted by smallholder farmers as an effective multiple-cropping system for food and feed production.

Mechanization and labor inputs for sub-Saharan Africa agricultural systems

In most agricultural systems in SSA, access to mechanization is extremely limited. The average number of tractors per 1000 ha in SSA is only about 11% of the developing world average (Houmy et al., 2013; Mrema et al., 2008). In 1961 SSA had more tractors than both Asia and the Near East, but in 2000, SSA had only about 221,000 tractors compared to 6 million in Asia and 1.7 million in the Near East (Sims et al., 2016). The majority of tractors in SSA are in just a few countries, with 28% in Nigeria and 32% in southern Africa (Mrema et al., 2008). There is no commercial tractor production in SSA outside of South Africa. The average farm size in SSA is 2 ha and 65%–80% of the land is estimated to be under manual production (Sims et al., 2016). With 500 million smallholder farmers producing 80% of the food worldwide (Sims et al., 2016), the sustainability and productivity of their agricultural systems are of tantamount importance.

A reliance on hand labor has made it possible and necessary for millions of African farmers to practice intercropping. While field preparation is done using tractors where they are available, most smallholders still plant, weed, and harvest their crops by hand. When pesticides are applied on smallholdings it is generally done with a backpack sprayer and can be targeted for intercropped rows. This lack of reliance on the tools used in monocultures means that there is little impedance to practicing the SCCS or intercropping in SSA. However, it is suggested by many scientists and policy makers that without advances in mechanization resources, SSA agricultural production will continue to decline and food scarcity will increase (Kienzle et al., 2013; Sims et al., 2016; Houmy et al., 2013). As innovation in mechanization becomes a focus of agricultural development in SSA, it should be done with forethought to conserving the traditional intercrop system and the many benefits brought forth by that production method. One way to do this is to design or utilize equipment that can easily fit into an intercropping system.

Currently several organizations are focusing on the potential of two-wheel tractors to improve agricultural productivity in SSA while still retaining the capacity to intercrop. The Farm Mechanization and

Conservation Agriculture for Sustainable Intensification (FACASI) project (http://facasi.act-africa.org/) led by the International Maize and Wheat Research Improvement Center (CIMMYT) has teamed with the University of Zimbabwe and other national agriculture research systems to test two-wheel tractors in an intercrop system (Fig. 7.1). The small tractors can pull one- or two-row planters for planting intercrop strips and can be outfitted with a small combine harvester unit that can reap, thresh, and bag grains and pulses.

The FACASI project also includes agronomic research to determine best planting time and row configurations for each intercrop component. In Fig. 7.2 the bean (*Phaseolus vulgaris*) crop was planted prior to the maize while in Fig. 7.3 the maize was planted a week before the bean intercrop. The success of various plant and row configurations is often dependent on the timeliness of rainfall. The project also utilizes east—west row orientation to allow the shorter crop to capture maximum sunlight.

In addition to two-wheeled tractors, small four-wheeled tractors may also be feasible for intercrop systems. The Feed the Future Soybean Innovation Lab (http://soybeaninnovationlab.illinois.edu/) promotes a

Figure 7.1 University of Zimbabwe researchers Dr. Raymond M. Nazare (right) and Tendai Chakandinakira (left) demonstrate a two-row tractor used in a FACASI intercropping study. *Photo courtesy of Kerry Clark.*

Figure 7.2 Maize planted a week after the bean intercrop. *Photo courtesy of Kerry Clark.*

Figure 7.3 Maize planted a week before the bean intercrop. *Photo courtesy of Kerry Clark.*

low horsepower four-wheel tractor that is an offshoot design of the Allis Chalmers Model G tractor made in the United States in the 1940s and 1950s. A four-wheel tractor gives the operator the opportunity to be seated while working, thereby reducing fatigue. A small four-wheel tractor can also be utilized in narrow crop strips used in intercropping although it has a higher initial cost than a two-wheel tractor. Most equipment produced in developed nations in the decades prior to farm consolidation and acreage expansion is small enough to be modified for use in intercropping in SSA. Many European and Asian equipment manufacturers also design and build with smaller acreages in mind. For many smallholder farmers, the cost of equipment, rather than feasibility of the equipment in the cropping system, is the main factor prohibiting use.

Potential benefits of solar corridor crop system to sub-Saharan Africa agricultural systems

While one of the primary benefits of the SCCS is presumed to be improved light penetration into the canopy, intercropping in SSA has been found to be beneficial for a large variety of reasons. These include complementary use of resources by intercropped species, efficient utilization of labor, suppression of weeds and insect pests, sequential harvesting, and better resistance to environmental stresses by diverse plant communities (Jalloh, 2001; Matusso et al., 2014). Perhaps the most important factor for cash-strapped smallholder farmers is the improvement in risk aversion when intercropping is used. Where drought and other environmental stresses are common, intercropping can provide stability through partial restoration of biodiversity, which can help reduce risk of total crop loss (Woomer et al., 2004; Machado, 2009; Himmelstein et al., 2017). This biodiversity can lead to reduced damage from insects, diseases, and weeds through deterrence mechanisms provided by companion crops. Maize intercropped with pigeon pea (*Cajanus cajan*) in Mozambique yielded eight times higher than monocropped maize, which often becomes infested with the parasitic weed species striga (*Striga asiatica*) (Rusinamhodzi et al., 2012). Research in Kenya found that striga is better controlled in intercropping compared to monocropping due to shading, higher humidity, and lower temperatures in the intercrop canopy (Oswald et al., 2002). The One Acre Fund, a farming development organization working in Kenya, experienced several strong hail events in 2014 and reported that legumes grown in monocrops were destroyed by the storms while those growing in an intercrop with maize were protected from damage by the maize leaves (Anonymous, 2015).

Physiological differences between crops can help maintain food security in times of climatic adversity. In central Mozambique, intercropped cowpea (*Vigna unguiculata*) was less affected than maize by long periods of low rainfall due to the shorter life cycle of cowpea and maturation coinciding with times of higher rainfall (Rusinamhodzi et al., 2012). In Kenya crop seasons with low maize yields in the intercrop were generally compensated by higher legume yields, with the reverse also being true. Because legumes and maize are sensitive to drought at different growth stages, intercropping helps spread the risk of unpredictable rainfall (Muchera-Muna et al., 2010). Additional but less visible benefits of mixed cropping systems include processes that occur in the rhizosphere of the intercropped plant roots. These processes may include improved nutrient recycling and availability, improvements in microclimatic conditions, better soil structural stability, increased microbiota habitat, and release and comingling of root exudates (Ndakidemi, 2006).

Although the SCCS focuses on radiation use efficiency (RUE) and light infiltration into the crop canopy, RUE may be of less importance to farmers in SSA due to the abundance of radiation available in the tropics and subtropics (Awal et al., 2006). Typically intercrops include a dominant cereal crop with erect architecture and bilateral leaf growth. The subordinate crop usually has planar leaves, allowing it to maximize radiation capture in the lower canopy (Awal et al., 2006). Several studies have found that decreased light penetration to the lower canopy caused by shading from the dominant crop leads to decreased intercrop yields of the lower species compared to monocrop yields. In a sorghum (*Sorghum bicolor*)/cowpea intercrop in Kenya (latitude 12°N), cowpea yields were reduced approximately 60% in the intercrop (Karanja et al., 2014). However, cowpea yield was also affected by row orientation, with east to west orientation producing an average of 15% higher grain yield than north–south–oriented rows. It was speculated that the east–west orientation allowed the cowpea to be shaded less by the sorghum and receive higher photosynthetic active radiation. Intercropped sorghum in the experiment was significantly shorter than monoculture sorghum due to resource competition from the cowpea and yielded less than sole-cropped sorghum. However, sorghum yields were higher when intercropped with two rows of cowpeas between each sorghum row than with alternating single rows of the two crops. North–south orientation of the intercrop and monocrop rows led to higher sorghum yields than east—west—oriented rows.

In a study in South Africa (latitude 29°06′S), sole maize was planted at 1.0 m row spacing while monocrop beans and beans intercropped with maize were planted at 0.5 m spacing (Tsubo et al., 2001). No significant differences were found in east—west versus north—south row orientation and no significant difference was found in RUE between sole maize and the intercrop. Productivity as measured by land equivalent ratio (LER) showed that the intercrop had an 8% yield advantage over the sole-cropped maize. LER is the total land area required in monocropping to produce the total yields obtained in an intercrop system (Tsubo et al., 2003). The maize in this experiment reduced intercropped bean yield by 90% compared to monocropped bean. In contrast to the SCCS in the United States where increased maize yield is a primary goal, intercropping in SSA places an important emphasis on the productivity of the understory crop as well as the dominant crop. Often the understory species is an important food security crop such as pigeon pea or cassava (*Manihot esculenta*) or important for human nutrition such as soybean, common bean, or groundnut (*Arachis hypogaea*), thus reduction in yields of the lower intercropped species can have repercussions beyond economic loss (Snapp et al., 2010). Yield of the intercropped bean and LER of the intercrop in the South African study were significantly higher where two rows of beans were grown between one-row of maize compared to alternating single rows of corn and bean.

Increasing the width of the intercrop strips also showed success in western Kenya, while the direction the rows were oriented showed no yield effects. Given the name MBILI, which means "two" in the Kiswahili language, a system of double maize rows planted at a 50 cm spacing separated by a 1-m intercropped legume strip resulted in 20% more light to the intercropped legume and a profit increase of \$69 ha^{-1} compared to alternating single rows of maize and legume (Woomer and Tungani, 2003). In a second study of MBILI with maize and common bean, light penetration to the understory increased by 54% over alternating row intercrop, the LER was 2.0, and maize yield increased 25%. Fertilizer use efficiency also improved due to more strategic placement of side-dress applications (Woomer et al., 2004).

In another South African study, double-rowed beans improved intercropping (latitude 23°53′S). When maize was planted at a low density the total intercrop yield decreased, and when beans were intercropped in double rows both maize and bean yields were higher than when bean and maize were in alternating single rows (Nthabiseng et al., 2015). The recommended planting density for the maize intercrop was 37,000 seeds ha^{-1} (Nthabiseng et al., 2015), which is near the maximum recommended monocrop maize population for dryland production in South Africa (Plessis, 2003). Low yields that many smallholder farmers experience in intercropping may be a result of low planting density. To improve outcomes for intercropping in SSA, researchers should determine the optimum planting populations of component crops.

In contrast to studies showing improved LER with wider legume strips, other research has found improved intercrop yields when the two component crops are grown in close proximity and planted in the same row. A within-row intercrop pattern in Ghana yielded significantly higher results than alternating one-row and two-row configurations for maize intercropped with cowpea, soybean, and groundnut. This was attributed to slightly better radiation capture by the within-row intercrop (Kermah et al., 2017). In Mozambique yield of maize in within-row intercropping with pigeon pea was higher than monocropped maize in 2 of 3 years and significantly greater than distinct row intercropped maize each year of the study (Rusinamhodzi et al., 2012). Pigeon pea has previously been found to have little adverse effect on intercropped maize because pigeon pea is a semiperennial that experiences a large part of its growth after maize has matured (Snapp et al., 2010). Although yields and resource sharing in within-row intercropping look promising, this production method would not lend itself well to mechanical cultivation and harvesting.

Resource availability and conservation in sub-Saharan Africa agricultural systems

Crop row configuration is not the only factor affecting light use in intercropping. Low RUE values in intercropping can be a result of a soil water deficit, demonstrating that in the dryland crop areas of SSA, water is the most limiting growth factor and radiation use is dependent on water use (Tsubo et al., 2003). In an ideal environment, radiation is the key driving force for plant growth (Tsubo et al., 2001), but in the semiarid regions of SSA, water and soil fertility are generally the key limiting factors for improving crop productivity (Karanja et al., 2014; Matusso et al., 2014). Less than 5% of the cultivated cropland in SSA is under irrigation, and the average fertilizer use in 2007 was 13 kg ha^{-1} compared to 208 kg ha^{-1} in

Asia and Latin America (Houmy et al., 2013). The addition of inorganic fertilizers to crop plants in SSA is very low due to access and affordability.

Several research and development organizations working on improving agricultural productivity in Africa promote intercropping as a sustainable system for farmers facing declining soil fertility and increased yield uncertainty due to climate change. In Zambia, CIMMYT (https://www.cimmyt.org/) leads an Africa RISING (https://africa-rising.net/) project that has shown that intercropping unfertilized maize with *Gliricidia sepium*, a leguminous tree, leads to a 66% improvement in maize yield over unfertilized maize grown alone. Lablab bean (*Lablab purpureus*) planted 3 weeks after maize also led to improved yield in the maize (Thierfelder et al., 2017). The One Acre Fund (https://oneacrefund.org/) uses the MBILI intercropping system designed by N2Africa researchers (http://www.n2africa.org/) to help farmers produce more sustainable and reliable yields. The World Agroforestry Centre (http://www.worldagroforestry.org/) encourages crop-tree intercropping and the World Bank–financed Agricultural Productivity Program for Southern Africa (http://projects.worldbank.org/) encourages female farmers to utilize intercropping and conservation agriculture. These are just a few of the many development programs that see the value in intercropping in agricultural production.

Intercrops are commonly used in low input–low output farming systems. In an experiment comparing a maize–bean intercrop to monocultures of both crop species at varying levels of resource availability, Temesgen et al. (2015) found no differences between cropping systems for maize yield at all resource levels and severely reduced yield for beans in the intercrop, which was improved at higher resource levels. Water capture as measured by crop evapotranspiration was not significantly different by cropping system, but water use efficiency was higher for the intercrop than for the sole bean and not significantly different from sole maize. Nitrogen use efficiency was also similar in the intercrop and the sole maize. LER was higher for the intercrop under low resource availability, indicating that intercropping may be more important for yield stability and risk aversion at low levels of inputs and resource availability. Kermah et al. (2017) also determined that LERs were higher for intercrops grown in lower fertility fields and more marginal environments than those grown under adequate water and fertility availability. Intercropping is not commonly practiced in northwestern Ethiopia, but due to soil fertility depletion from continuous maize cropping, researchers are investigating the benefits of mixed cropping. The intercrop system of maize and common bean produced 20% more yield than the sole maize system and the economics of maize production was improved through intercropping at all levels of added nitrogen (Alemayehu et al., 2017). In many low-input smallholder systems, animal manure may be the primary source of nutrients and can help complement the nitrogen produced by legumes in an intercrop systems. Figs. 7.4 and 7.5 show the beneficial effects of added cattle manure on plant growth in a corn–soybean intercrop in Mozambique.

In developed nations, mixed-stand cropping is routinely practiced in pastures for animal forage because both scientists and farmers recognize that it contributes to higher productivity per unit of land and leads to healthier plant ecosystems, but this same approach is not usually practiced when it comes to our grain crop systems (Belshaw, 1979). By planting in pure stands, we are limiting our intercrop complementary effects, but we are maximizing our ease of production. In monocropping, we can easily control mechanized operations and we can focus on increased output of single commodity crops. Currently in SSA, food production is usually more emphasized than commodity production, thus intercropping may be

Figure 7.4 Maize–soybean intercrops on a farm in Mozambique planted on soil enriched with cattle manure. *Photo courtesy of Kerry Clark.*

Figure 7.5 Maize–soybean intercrops on a farm in Mozambique planted on soil with no amendments. This is located less than 10 meters from the cropping area shown in Fig. 7.4. *Photo courtesy of Kerry Clark.*

more socially acceptable. While SSA farmers need to make profits as much as developed-nation farmers, smallholders in SSA also need to produce the crops that their families depend on for survival. Intercropping allows them to match a diverse human diet with a diverse cropping system. What a consumer in the United States gets from a grocery store, smallholder farmers in SSA usually must grow themselves or go without. Agricultural biodiversity is an important component of food sovereignty (Tittonell, 2014).

Socioeconomic aspects of intercropping as a food production system in sub-Saharan Africa

As human population is expected to reach a new high of 9 billion by 2050, there is much discussion of how the world will increase food production by at least 50%. Many policy makers and agriculturalists believe that SSA will have to drastically increase its crop productivity to meet future food production goals (Tilman, 1999; Horlings and Marsden, 2011). There are several paths that SSA can take to do this. The path that the United States has followed, including high inputs of fertilizer, pesticides, and large equipment is fraught with issues of unsustainability. Although yields increase when going from no fertilization to fertilization, the efficiency of fertilizer use is decreasing over time, possibly due to maximum yield potential being reached in improved crop varieties, or possibly due to a breakdown of the productive base of agriculture in heavily monocropped systems that lack ecological resiliency (Altieri, 1998). From 2000 to 2007, fertilizer prices increased 113% and problems associated with high fertilizer use including surface and groundwater pollution, soil acidification, and ammonia volatilization have all increased and added to ongoing environmental degradation (Machado, 2009; Altieri, 1998). Synthetic pesticide use in technology-based agricultural systems has also increased drastically, but due to pesticide resistance and outbreaks of secondary pests, yields are still threatened by insects, pathogens, and weeds (Machado, 2009). Because our monoculture systems can no longer provide adequate biological pest suppression, approximately $40 billion is spent annually in the United States on pest control to save approximately $16 billion in crops (Altieri, 1998). In California, pesticide use increased from 161 to 212 million pounds of active ingredient from 1991 to 1995, although the statewide crop acreage did not increase (Altieri, 1998).

An alternative path for SSA agriculture is one of ecological intensification in which emphasis is placed on diversity, efficiency, recycling, and regulation, which are the pillars of agroecology (Tittonell, 2014). Agroecology relies primarily on diversification of species or functional traits across time and space. In ecological intensification, crop yields depend primarily on internal resources such as natural rainfall, biological control of pests, and nutrient recycling of organic matter (Altieri, 1998). Either rotations contain many diverse species, or diverse species can be interplanted at the field level. Because intercropping preserves functional biodiversity, it can be an important aspect of ecological intensification. It has also been established that small and diversified farms show greater productivity per land area than do large monocultures (Tscharntke et al., 2012).

However, there are factors at play that can influence the agricultural development paths taken in various areas of SSA. For example, land tenure issues and lack of private land ownership in many SSA countries can lead to land grabs that force out smallholder farmers. According to Horlings and Marsden (2011), investors expressed an interest in 32 million ha of land in SSA in 2008. Approximately 60 million ha have been purchased in recent years in SSA, primarily by multinational corporations from Brazil, Russia, India, China, and South Africa (Sharife, 2012). As this land moves from control by smallholder farmers to large corporations, agricultural production becomes commodity oriented, inequalities are deepened, and food security may be further threatened (Sharife, 2012; Batterbury and Ndi, 2018). These changes in land ownership may force agriculture in SSA to follow the same path of input reliance in a low diversity agroecosystem that many developed countries have followed.

Additionally, although intercropping has a long history in SSA, its use has decreased in many areas. In Zimbabwe, smallholders were encouraged by the extension service to follow the path of large commercial farmers and use monocropping. As a result, intercrops are rarely found in Zimbabwean agriculture and yields have declined in the past decades (P. Mapfumo, personal communication, March 6, 2018). In Zambia, 72% of farmers surveyed in the Magoye area did not use any intercropping due to pressure to produce marketable surplus from a monocrop system. Women and uneducated farmers were the least likely people surveyed to employ intercropping (Chijikwa, 2012). Visits to farms in Mozambique will reveal multiple crops growing in intercropped fields, but farm visits in Ghana will most likely show a majority of fields in monocrop production.

Summary

For intercropping to remain viable in SSA or to be readopted as an SCCS in technology-based agricultural systems, there needs to be a rethinking of planting, fertilization, spraying, and harvesting of multiple crops in the same field (Machado, 2009). In SSA, research needs to build upon indigenous knowledge (Tittonell, 2014). Intercropping has historically been seen as a primitive system that will eventually give way to monocropping as access to technology increases, but that may not be the best way forward. Intercropping should not be viewed as a step backward, but instead, its useful aspects should be developed for modern agriculture. Potential intercropping interventions and areas for research include crop variety development under intercrop conditions, determination of appropriate planting times for component crops, fertilization rates and regimes, appropriate planting patterns, and developing appropriate mechanization for planting, weeding, and harvest. In the United States, the SCCS and similar intercropping practices can easily fit into organic systems (as equipment permits) due to the absence of synthetic pesticide application. In conventional agriculture, chemical application is much more challenging in a multicrop system than in a monocrop system, but it is still feasible. In large-scale conventional agriculture, even the use of wider intercrop strips that allow the passage of modern equipment will help improve interfield biodiversity, which may contribute to better biological control of insects and diseases, better light penetration into the crop canopy, and improved yields. The split-row combine in Fig. 7.6, built by the research equipment company ALMACO (Nevada, IA, USA), can harvest and segregate seeds of two different crops planted side by side. Similar equipment modifications may be used to increase the feasibility of SCCS or intercropping in technology-based production systems.

Intercropping is the global practice of the SCCS and is a pivotal production system for African smallholder farmers. As agriculture development occurs in SSA and other developing nations, both society and the environment will benefit by retaining as much biodiversity as possible by modifying our expectations and vision of what a successful agricultural system looks like. In the United States and Europe, we once successfully fed the population with low-input, highly diversified farms. In the older, more degraded soils of SSA, that goal is more difficult without some modern agricultural inputs. Also because population has increased and shifted from rural to urban, we are dependent on increased agricultural productivity per land area. Intercropping was abandoned in the United States after the industrialization of agriculture began after World War II, but that abandonment may have been more a matter of convenience and

Figure 7.6 A split combine capable of harvesting two different crops at the same time. *Built by ALMACO (Nevada, IA, USA). Photo courtesy of ALMACO (https://www.almaco.com/).*

not because it was inherently incompatible with modern agriculture (Horwith, 1985). The benefits of reintegrating intercropping systems, such as the SCCS, back into modern agricultural practice will very likely outweigh the loss of convenience from abandoning the current monoculture system (Horwith, 1985; Horlings and Marsden, 2011; Tilman, 1999; Altieri, 1998). In both developed and developing nations, policy makers, scientists, farmers, and consumers should give serious thought and action to combining the best aspects of these systems to produce a new and truly sustainable way to provide global food security and improved yields, ecosystem services, and environmental protection.

References

Alemayehu, A., Tamado, T., Nigussie, D., Yigzaw, D., Kinde, T., Wortmann, C.S., 2017. Maize–common bean intercropping to optimize maize-based crop production. J. Agric. Sci. 155 (7), 1124–1136.

Altieri, M.A., 1998. Ecological impacts of industrial agriculture and the possibilities for truly sustainable farming. Monthly Rev. 50 (3), 60.

Anonymous, 2015. Maize-legume intercropping. One Acre Fund 2014 Phase 1 and 2 Trial Report. <https://oneacrefund.org/insights-library/?page = 2&category = 16&category = 14> (accessed 20.05.18.).

Awal, M.A., Koshi, H., Ikeda, T., 2006. Radiation interception and use by maize/peanut intercrop canopy. Agric. Forest Meteorol. 139 (1–2), 74–83.

Batterbury, S., Ndi, F., 2018. Land-grabbing in Africa. The Routledge Handbook of African Development. Routledge International Handbooks. Routledge, London, pp. 573–582.

Belshaw, D., 1979. Taking indigenous technology seriously: the case of inter-cropping techniques in East Africa. IDS Bull. 10 (2), 24–27.

Chijikwa, M., 2012. Factors affecting a farmer's willingness to adopt intercropping: a case study of farmers in Magoye, Zambia. <https://www.icac.org/wp-content/uploads/2012/09/Zambia_Intercropping1.pdf> (accessed 30.05.18.).

Christensen, R.P., Hendrix, W.E., Stevens, R.D., 1964. How the United States Improved Its Agriculture. United States Department of Agriculture (USDA) Economic Research Service. Development of Trade Analysis Division.

Cook, R.J., Weller, D.M., 2004. In defense of crop monoculture. In: Proceedings of the 4th International Crop Science Congress, vol. 26, pp. 1–11.

Himmelstein, J., Ares, A., Gallagher, D., Myers, J., 2017. A meta-analysis of intercropping in Africa: impacts on crop yield, farmer income, and integrated pest management effects. Int. J. Agric. Sustain. 15 (1), 1–10.

Horlings, L.G., Marsden, T.K., 2011. Towards the real green revolution? Exploring the conceptual dimensions of a new ecological modernisation of agriculture that could 'feed the world'. Global Environ. Change 21 (2), 441–452.

Horwith, B., 1985. A role for intercropping in modern agriculture. BioScience 35 (5), 286–291.

Houmy, K., Clarke, L.J., Ashburner, J.E., Kienzle, J., 2013. Agricultural Mechanization in Sub Saharan Africa: Guidelines for Preparing a Strategy. United Nation Food and Agriculture Organization, <http://www.fao.org/3/a-i3349e.pdf> (accessed 20.05.18.).

Jalloh, A., 2001. Promoting Appropriate Intercropping Technologies for Sustainable Agricultural Production in Africa: A Farmer Centered Approach. Institute of Agricultural Research, Freetown, Sierra Leone.

Karanja, S.M.S.M., Kibe, A.M., Karogo, P.N., Mwangi, M., 2014. Effects of intercrop population density and row orientation on growth and yields of sorghum-cowpea cropping systems in semi-arid rongai, Kenya. J. Agric. Sci. 6 (5), 34.

Kermah, M., Franke, A.C., Adjei-Nsiah, S., Ahiabor, B.D., Abaidoo, R.C., Giller, K.E., 2017. Maize–grain legume intercropping for enhanced resource use efficiency and crop productivity in the Guinea savanna of northern Ghana. Field Crop. Res. 213, 38–50.

Kienzle, J., Ashburner, J.E., Sims, B.G., 2013. Mechanization for Rural Development. United Nation Food and Agriculture Organization, <http://www.fao.org/docrep/018/i3259e/i3259e.pdf> (accessed 20.05.18.).

Knörzer, H., Graeff-Hönninger, S., Guo, B., Wang, P., Claupein, W., 2009. The rediscovery of intercropping in China: a traditional cropping system for future Chinese agriculture—a review. Climate Change, Intercropping, Pest Control and Beneficial Microorganisms. Springer, Netherlands, pp. 13–44.

Kremer, R.J., Deichman, C.L., 2014. Introduction: the solar corridor concept. Agron. J. 106 (5), 1817–1819.

MacDonald, 2013. Crop consolidation and the future of the family farm. Amber Waves: The Economics of Food, Farming, Natural Resources, and Rural America. United States Department of Agriculture (USDA) Economic Research Service, <https://www.ers.usda.gov/amber-waves/2013/september/cropland-consolidation-and-the-future-of-family-farms/> (accessed 20.05.18.).

Machado, S., 2009. Does intercropping have a role in modern agriculture? J. Soil Water Conserv. 64 (2), 55A–57A.

Matusso, J.M.M., Mugwe, J.N., Mucheru-Muna, M., 2014. Potential role of cereal-legume intercropping systems in integrated soil fertility management in smallholder farming systems of Sub-Saharan Africa. Res. J. Agric. Environ. Manage. 3 (3), 162–174.

Mrema, G.C., Baker, D., Kahan, D., 2008. Agricultural Mechanization in sub-Saharan Africa: Time for a New Look. United Nation Food and Agriculture Organization, <http://www.fao.org/3/a-i0219e.pdf> (accessed 20.05.18.).

Mucheru-Muna, M., Pypers, P., Mugendi, D., Kung'u, J., Mugwe, J., Merckx, R., et al., 2010. A staggered maize–legume intercrop arrangement robustly increases crop yields and economic returns in the highlands of Central Kenya. Field Crop. Res. 115 (2), 132–139.

Ndakidemi, P.A., 2006. Manipulating legume/cereal mixtures to optimize the above and below ground interactions in the traditional African cropping systems. Afr. J. Biotechnol. 5 (25), 2526–2533.

Nelson, K.A., 2014. Corn yield response to the solar corridor in upstate Missouri. Agron. J. 106 (5), 1847–1852.

Nthabiseng, T.R., Irvine, K.M., Moshibudi, P.M., 2015. Response of a maize or dry bean intercrop to maize density and dry bean arrangement under rainfed conditions. Int. J. Agron. Agric. Res. (IJAAR) 6 (6), 18–29. <http://www.innspub.net> (accessed 20.05.18.).

Oswald, A., Ransom, J.K., Kroschel, J., Sauerborn, J., 2002. Intercropping controls Striga in maize based farming systems. Crop Protect. 21 (5), 367–374.

Plessis, J., 2003. Maize Production. Department of Agriculture, Pretoria, South Africa, <http://www.arc.agric.za/arc-gci/Fact%20Sheets%20Library/Maize%20Production.pdf> (accessed 30.05.18.).

Rusinamhodzi, L., Corbeels, M., Nyamangara, J., Giller, K.E., 2012. Maize–grain legume intercropping is an attractive option for ecological intensification that reduces climatic risk for smallholder farmers in central Mozambique. Field Crop. Res. 136, 12–22.

Sharife, K., 2012. Land grabbing in Africa. <http://commons.ln.edu.hk/southsouthforum/2012/s4/2/> (accessed 30.05.18.).

Sims, B., Hilmi, M., Kienzle, J., 2016. Agricultural Mechanization. A Key Input for sub-Saharan African Smallholders. United Nation Food and Agriculture Organization, <http://www.fao.org/documents/card/en/c/51868c45-4ed7-4733-94ae-751b2e5589f7/>. (accessed 20.05.18.).

Snapp, S.S., Blackie, M.J., Gilbert, R.A., Bezner-Kerr, R., Kanyama-Phiri, G.Y., 2010. Biodiversity can support a greener revolution in Africa. Proc. Natl. Acad. Sci. U.S.A. 107 (48), 20840–20845.

Temesgen, A., Fukai, S., Rodriguez, D., 2015. As the level of crop productivity increases: is there a role for intercropping in smallholder agriculture. Field Crop. Res. 180, 155–166.

Thierfelder, C., Heinrich, G., Lewis, D., Mwila, M., Bekunda, M., 2017. Intercropping strategies under conservation agriculture: Africa RISING science, innovations and technologies with scaling potential from ESA-Zambia. In: Poster Prepared for the Africa RISING Science for Impact Workshop, Dar es Salaam, January 17–19, 2017, CIMMYT, Mexico City, Mexico.

Tilman, D., 1999. Global environmental impacts of agricultural expansion: the need for sustainable and efficient practices. Proc. Natl. Acad. Sci. U.S.A. 96 (11), 5995–6000.

Tittonell, P., 2014. Ecological intensification of agriculture—sustainable by nature. Curr. Opin. Environ. Sustain. 8, 53–61.

Tscharntke, T., Clough, Y., Wanger, T.C., Jackson, L., Motzke, I., Perfecto, I., et al., 2012. Global food security, biodiversity conservation and the future of agricultural intensification. Biol. Conserv. 151 (1), 53–59.

Tsubo, M., Walker, S., Mukhala, E., 2001. Comparisons of radiation use efficiency of mono-/inter-cropping systems with different row orientations. Field Crop. Res. 71 (1), 17–29.

Tsubo, M., Mukhala, E., Ogindo, H.O., Walker, S., 2003. Productivity of maize-bean intercropping in a semi-arid region of South Africa. Water SA 29 (4), 381–388.

Woomer, P.L., Tungani, J.O., 2003. Light availability within an innovative maize-legume intercropping system in Western Kenya. In: African Crop Science Conference Proceedings, vol. 6, pp. 42–46.

Woomer, P.L., Lan'gat, M., Tungani, J.O., 2004. Innovative maize-legume intercropping results in above-and below-ground competitive advantages for understorey legumes. West Afr. J. Appl. Ecol. 6 (1), 85–94.

CHAPTER 8

Vision

C. LeRoy Deichman[1] and Robert J. Kremer[2]
[1]Deichman Consulting, Shelbyville, MO, United States
[2]University of Missouri, Columbia, MO, United States

Contents

Introduction

The challenge for agriculture in the 21st century is to implement more sustainable farming systems that are economically viable, accommodate changing technologies and climate, and provide sufficient, high-quality food to meet global needs (Johnson et al., 2017; Lehman et al., 2015a). This suggests that food and fiber production that results in an increased carbon footprint due to increased inputs of fuel, fertilizers, and pesticides under current farming systems must be shifted toward more sustainable management for food production that is economically viable for farmers and rural communities and promotes ecological processes for better nutrient management and reverses the degradation of soil health and environmental quality. A vision for the solar corridor crop system (SCCS) as an alternative management practice that maintains crop yields while minimizing external inputs to ensure economic and environmental sustainability was predicated on an opinion piece by Kremer and Deichman (2016) that questioned the sustainability of current industrial-scale crop production based on monocultures of commodity crops with indiscriminate inputs of

The Solar Corridor Crop System
DOI: https://doi.org/10.1016/B978-0-12-814792-4.00008-5

synthetic fertilizers and pesticides that inherently reduce environmental quality. Our vision, in due time and with continued successful development, foresees utilization of the SCCS as necessary to enable a world in which:

1. We can potentially double the world corn production with full implementation of the SCCS (Deichman et al., 2016).
2. No beginning farmer (BF) has to accept the pervasive narrative that there is no way to successfully begin farming without megabucks in hand.
3. No competent hard-working family has to be forced off of the farm by the mortgage holder simply because the family farm, which was once large enough, is now too small.
4. No well-raised farm youth who wants to stay on or return to the successful family-owned farm has to abandon that understandable desire due to potential insufficient net operating profits.
5. We can reverse the exodus from (and of) the small acreage family farm (SFF).
6. Rural towns currently in long-term decline, where corn is successfully produced, can potentially reverse that long-term trend.
7. A renewed hope of an unprecedented socioeconomic (community) recovery, wherever corn and corn-compatible crops can be grown, is justified.
8. We can restore degraded lands by rebuilding soil biology and other soil properties and reducing current erosion levels of our vital topsoil.
9. We are mitigating our rampant environmental degradation via our abuse of our current carbon cycle (CC) and disruption of vital ecological services.
10. We are no longer addicted to oil.
11. We are using significant amounts of the infinite quantities of earth's atmospheric-sourced CO_2 cycling in the current CC to replace nonrenewable fuels that source their carbon from the CC that cycled when dinosaurs roamed the earth.
12. We can produce enough additional food to end starvation in food-deficit regions around the globe.

Our vision, in summary, offers a unique opportunity for enhancing soil health/sustainability, the viability of the family farm, rural communities, air and water quality for future generations, and the unleashing of earth's long-term potential bioproductivity.

This chapter, at a minimum, intends to demonstrate the basic mechanics of one way each of above objectives could be achieved. This concluding chapter of our seminal volume, offers our founding and sustained vision of what could be achieved with full deployment of the SCCS.

Crop productivity and the family farm

Based on the lack of information on the idea prior to our research on the solar corridor concept, the nearly inconceivable magnitude of hybrid-specific yield response per our vision offered here, it is appropriate to preface this chapter by summarizing the well-documented yields disclosed in previous chapters as follows.

With three high-yielding non-GMO (genetically-modified organism) hybrids from Deichman's 1990 preliminary screening trials having been identified and/or confirmed as being potentially responsive to the solar corridor, we consistently produced as much or more corn from half as many twin rows at 60-, 72-, and 76-in. row widths as state-of-the-art, best management practices controls produced in conventional 30-, 36-, and 38-in. row widths (see Figs. 2.1–2.6 and Table 2.1). These results were obtained across seven different Corn Belt soils under the extreme drought conditions of 1991 or under more typical good growing seasons in subsequent years (Deichman, 2019). This produced lowest population "control" yields of up to 50 bu $acre^{-1}$ less than trend line yields in the drought year to solar corridor treatment yields of 50 bu above trend line yields (Nielsen, 2017) in the more typical growing seasons thereby demonstrating an amazingly similar (same) response over a wide yield range (100 bu). In independent peer review yield results, Nelson (2014, in Table 2) duplicated those results with a GMO hybrid, on a different soil and under different growing conditions without benefit of a hybrid that met the preliminary hybrid screening criterion that were met by hybrids B, C, and D in our seminal and preliminary hybrid screening trials (Deichman, 2019). When citing Nelson's research, which replicated the response of our seminal research, using 60-in. row widths, we typically indicate 60- instead of 60- to 72- or 76-in. rows (Fig. 8.1).

The SCCS corn/winter grain/stubble/cover crop, as illustrated in Fig. 2.8, enables the biosynthesis and sequestration of more atmospheric-sourced CO_2 than any current monocrop system known to these authors. It is obvious to all who are knowledgeable of the SCCS that it enables a

Figure 8.1 Field-scale test strips of corn planted in 60-in. (152-cm) row widths in Illinois. *Photo courtesy Bob Recker, cedarvalleyinnovation.com.*

significant reduction of soil erosion, residual nutrient, and pesticide losses from our finite cropland that, without the SCCS, would have been thrust and/or gravitated into potential drinking water sources. This is a serious problem that is irresponsibly ignored or dismissed by many who should know better. Each SCCS design should recognize this SCCS-enabled opportunity and plan accordingly. Even though the SCCS is much easier to deploy in large-scale crop production when soybeans are the "main crop" and winter wheat or other winter grain and stubble crop is the corridor floor crop (as is the case with many other floor crops), its promising potential has yet to be developed. However, considering its ease of deployment as another main crop in the SCCS, we envision an expedited interest and subsequent development of this (soybean/wheat) crop sequence in the SCCS. Our vision predicts the continuation and improvement of performance capabilities of the properly deployed SCCS along with further increases in the improvement of wheat yield response to the solar corridor as well as most other agronomic crops, including soybeans as the main crop, as the many potential site-specific synergies are identified, effectively addressed, and actively deployed.

Appropriate adaptations of the SCCS to fit the various specific situations that enhance the intergenerational sustainability of the SFF and sustainable viability of the BF, which enables a subsequent reversal of the exodus of (and/or from) the SFF and small towns in rural America, which enables subsequently earned distributive prosperity. One such adaptation would substitute appropriate high-value spring (i.e., July 4th sweet corn) and fall community-supported agriculture (CSA) crops for wheat, at an appropriate distance on the end of each wheat swath with the first and last full-length rows devoted to perennial crops of strawberries, asparagus, rhubarb, flowers (for holidays from Easter to Memorial Day), etc. The contracted CSA market, school lunch, food bank, USDA-designated food desert status and available labor would most precisely determine "appropriate," as referenced above. The SCCS enables the gainful employment of more people working the land. In fact, more people on the land are required to optimize the productivity increases enabled by the SCCS.

The current land (L) to people capable (CP) of farming the land ratio (L:cp) in the United States has completely reversed since the colonists broke through the Cumberland Gap. The challenge then was how to farm all of that newly discovered farmland, represented by the ratio L:CP. Now the ratio is CP:l yet we are still using the crop production paradigm that effectively worked to conquer that daunting problem. The larger and even larger farm equipment has driven the production paradigm during the past 150–200 years. (The popular standard answer of farm size typically satisfies the casual farm status inquiry). The SCCS can put us back onto the track of prioritizing the sustainable productivity of our limited land resource base and subsequently revitalizing rural communities instead of the inherently destructive race for more land by fewer and fewer farmers. The number of farm operations are decreasing. Too many competent farm-raised youth who want to stay on the farm and potential farmers who want to begin farming are being counseled against it.

SCCS can reverse this unfortunate trend.

1. It is the authors' belief that many competently operated SFFs have been unnecessarily forced out of existence under the auspices of being too small. This has caused disastrous personal and societal consequences to people, families, and rural towns with such an unimaginable impact that it defies apt description. We wish we would have had the SCCS ready for those who were so unjustifiably denied.
2. The authors believe that every capable industrious child with a burning desire, positive mental attitude, and parental/educational guidance from the youngest ages can stay on the land without having to push their neighbors off their land and putting over-width farm equipment on the roads.
3. The authors believe that we should appropriately incentivize the BF. Many options are available to fit the diverse personal farm situations that address all three issues. For example, if you are near one of the many USDA-designated food deserts, you have a logical choice.

Integrate CSA crops, described above, into SCCS-based farm operations as family and other labor availability permits.

Livestock operations that fit into the farm operation offer maximum synergies. Other options challenge your most insightful, individual personalized thoughts. Hopefully, many readers of this seminal volume on the SCCS will be active participants in developing these base-broadening and subsequent commercial deployment objectives.

Specific entities who could uniquely benefit from the SCCS are as follows:

BF

SFF

Third World farmer with 2–3 ha of land

The farm-raised youth who wants to stay on the farm

The small rural town in long-term demise in the US Corn Belt

Economically deprived communities wherever crop production can flourish

Outstanding corn grower who could handle more land but doesn't have more land.

Typical opportunities

Crop choices: The SCCS provides opportunities to try several crops as the primary component in a two-crop system, which is illustrated below.

Main crop: Corn

Production goal: 100 bu wheat/robust stubble forage/continuous cover crop on the solar corridor floor half of the field - 400 bu corn on the corn half of the field (Deichman et al., 2016).

Unique yield source: The floor crops, planted as shown in Fig. 2.8, have the benefit of unlimited access to incident sunlight most of the spring via no competition by the main crop early and little competition until clover is ready to bloom and wheat is reaching maturity. As a matter of fact, depending on the cultivars and growing season, the stubble clover can be in bloom and the winter grain mature before the corn reaches incident sunlight dominance.

Major constraints: Equipment; floor crop cultivars haven't been sufficiently identified and researched.

Main crop: Soybeans

Production goal: 105 bu wheat/100 bu soybeans/robust stubble forage/continuous cover crop.

Unique yield source: The floor crops, planted as shown in Fig. 2.8, have a benefit of unlimited access to incident sunlight all spring via no competition by the main crop. The winter grain and stubble crops can do even better than described as for corn, above (Remember: main crop and floor crop each have half of the spatial field area.).

Major advantage: No new equipment required for winter grain harvest, eliminating an obstacle to not actively consider the SCCS.

Major constraints: Hasn't been sufficiently researched.

Main crop: Corn

Floor crop: Soybeans

Constraints: Research to date has not been successful. Despite vigorous early growth in this system, soybean performed poorly in mid- to late-season (Smith, 2000), thus considerable research is still required. We may have been most successful using hybrid D effectively and in consultation with a collaborating soybean physiology agronomist, but the discontinuance of hybrid D ended our efforts toward this goal.

The bright spot: Successfully operating equipment for all commercial operations was built and used for the short duration of our research. In this case the agricultural engineering team excelled, the agronomy team didn't. The soybean combine, surprising most of us, effectively harvested the beans between the wide corn rows.

All other crops:

Constraints: Preliminary trials with grain sorghum (Kremer et al., 2017) and sunflower (Mullally, 2011) as main crops have been conducted; however, research is insufficient to nonexistent.

For the economically disadvantaged SFF with a competent, ambitious growing family, the SCCS provides a sustainable opportunity to prosper.

For the competent, ambitious farmer in Chapter 12 (U.S. Bankruptcy) the SCCS provides a successful recovery plan.

For the well-prepared, ambitious BF the SCCS provides a sustainable opportunity to begin and navigate the long route to well-earned socio-economic success, that is, raise a couple of next-generation farmers and pay off the mortgage.

Impacts on the rural community

The SFF near a moderately populated small town in the US Corn Belt that has lost its grocery store(s) has an opportunity to revive and sustain the rural community. SFF, as specified above, represents a minimum threshold of about 20 acres to a maximum of about 640 acres of contiguous acres within a section without requiring expensive road transport with field equipment. Using CSA floor crops in the properly designed, developed, and deployed SCCS can enable the family farm to prosper and be sustained for generations, regardless of any financial or social institution mandates that they get bigger or get out. The "beginning farmer," as officially defined elsewhere, can successfully purchase a farm and raise a family.

Here we suggest some appropriate crop choice examples. For the farm that has erosion potential on all acres, the farmer may use corn as the primary crop with a winter small grain as the floor crop followed by stubble and subsequent cover crop that could best suit a dairy or beef operation. For the farm that has very low erosion potential, the farmer should use corn as the primary crop with a winter small grain or oilseed crop as the floor crop and stubble/cover crop on all but a selected distance from the end of the corridor, which is planted to appropriate early spring and late fall CSA crops, eg., peas, turnips, respectively.

Essential enabling elements (3E) include equipment purchase savings by acquiring appropriate small-scale equipment or even foregoing the purchase of a combine, not requiring "trucking" components for road transport to be manufactured into the equipment, operating savings by eliminating road time, wear and tear on the roads, equipment, dual-purpose rubber tires, etc. The assessment of the negative social, including the often fatal impacts of slow moving, oversize farm equipment on the public roads is beyond the scope of this volume. If we accept the widely held conclusions that larger and larger farms are necessary and inevitable to modern development and accordingly plug that into a mathematical assumption, it typically sources faulty assumptions that could have far-reaching impacts.

Based on the various possibilities with properly deployed SCCS, we envision putting the "culture" back in agriculture by enabling every uniquely capable young farm family to begin, grow, and sustain the family farm. The family farmer can determine the optimum crop choices for his/her site-specific SCCS. We believe very desirable outcomes at the rural community level can be realized including greater sustainable productivity per spatial unit; greater number of profitable sustainable farms; greater source of local, identity-preserved foods; more farmers on the land; stronger, more resilient, and sustainable rural communities; and a socioeconomic community structure and platform that enables every farm youth who puts in the effort, learns the requisite lessons growing up on the farm and who wants to farm; or to have a sustainable economically viable place on the farm.

The terminal misery of starvation on one hand and the myriad of destructive forces currently imposed on the traditional family farm and small rural towns in highly developed monocrop agricultural areas of the United States are beyond the scope of the authors' ability to effectively convey.

The opportunities for biofuel production with the solar corridor crop system

For the community that is seeking help to recover from its addiction to oil, the SCCS has produced enough corn to produce over 1000 gallons of anhydrous ethanol per acre from the corn starch component. One production model is outlined below.

Offer an opportunity to every farmer within a specified distance from the biofuels plant by the following:

1. Investing X bu or X$ per bu of pre-SCCS corn produced on the farm.
2. Deploying your site-specific SCCS on all of your corn acres.
3. Taking back biodiesel, E-85, E-85-B, and/or hydrogen for fuel cells, respectively, if or as approved, the markets indicate potential viability and the biofuel plant is capable of producing the requisite fuels.
4. Taking back wet distillers grain/gluten feeds or 60% protein dry gluten meal.
5. Working in the local producer- and employee-owned biofuels plant or new paradigm distribution network.

The "food and fuel paradigm" enables a reduction of CO_2 and an increase of oxygen in the air that we and future generations will breathe. The maximum utilization deployment of two 180 degree paradigm shifts would reduce atmospheric CO_2 from the current CO_2 by (1) capturing more atmospheric CO_2 in the increased production of crops and diverting all of the "net" crop increase into responsible biofuels production that displaces carbon-sourced fossil fuels from the current CO_2 that cycled when dinosaurs roamed earth. One of the reasons we specify "net" crop increase is some of the increased atmospheric CO_2 results in increased sequestration directly into the soil rhizosphere as one component of the subsequent improvement in soil health. Deichman et al. (2016) offers a seminal paper illustrating how 60–90 billion annual gallons of anhydrous ethanol equivalent biofuels could be produced in the US heartland from corn only; and (2) using the biofuels in the community from which it is sourced.

This means in the heart of the Corn Belt we would not be required to transport corn more than a short predetermined distance from its production to the biofuels plant. Currently, billions of bushels produced in Kansas, Nebraska, South Dakota, North Dakota, Iowa, Missouri, and Western Minnesota find their way to distant gulf or lake ports that in this paradigm shift are only transported a short distance to a community biofuels plant. We know that it is a long way from the aforementioned western Corn Belt areas to any navigable waters and more trucks on the road are unnecessarily relied on. With the socioeconomic makeup of the heart of the heartland in the US Corn Belt, the proven economics of current biofuels plants, the existence of 20 million flex-fuel vehicles on the road and millions produced each year, the awareness of needed markets by the corn-, soybean-, and wheat-centered producer organizations and the potential clout of the major farm organizations and the agricultural industry, little imagination is needed to know that a major part of the annually produced 60–90 billion gallons could be consumed without traveling more than a few miles from the biofuels plant. If we compare this simple short transportation requirement to the current transportation paradigm of shipping corn to overseas destinations and importing oil from overseas ports, we can see another example that complements the above paradigm and offers an easily overlooked or new dimension to the term "net" energy production. Multiple benefits beyond the singular direct displacement of fossil-sourced natural gas, diesel, and gasoline with the biofuels that would displace them are readily visualized. These biofuels include bioethanol for flex-fuel autos, biobutanol for "drop in fuels," biodiesel for compression ignition internal combustion engines, and biogas to replace natural and LP gas. The key is simply to consider it, but with our current addiction to oil (G.W. Bush, 2006 SOTU), such considerations are circumvented. Thus simple consideration is a step forward. If E-85-B were approved, gasoline could be completely replaced by biofuels instead of the displacement term currently applicable.

Smaller, higher compression spark ignition internal combustion engines could hasten the efficiency of ethanol fuels or, I wonder, what the efficiency of the higher octane ethanol would be when engines were designed to maximize the unique combustion properties of ethanol fuels. We know of no current US-built automobile that was designed to utilize the unique combustion properties of ethanol. E-85 flex-fuel automobiles are currently using engines designed for gasoline but are made compatible with ethanol blends up to 85%. It is understandable why compression-ignition diesel engines are designed for the unique combustion properties of diesel and gasoline engines for the unique production properties of gasoline. We speculate that a higher compression engine of 12 or 13:1 compression ratio and a more rectangular bore stroke configurations would (for the consumer) be worth further investigation before valid mileage comparisons between gasoline and ethanol can be made. (The readers of this book are aware of the fallacy of comparing apples and oranges.) That is, however, exactly what is done when comparing the miles per gallons of gasoline in engines designed for the unique combustion properties of gasoline anhydrous ethanol.

Farmers using the SCCS could take advantage of the 1000-gallon program whereby enough corn is produced for generating 1000 gallons of anhydrous ethanol, or equivalent alcohol fuels, from corn starch only. The plan provides for additional biodiesel, as needed, to plant and harvest the crops, and for biogas to heat and light the farm house, shop, and barn from further downstream processes.[1]

Impacts on international agriculture and global hunger

The SCCS enables the production of enough additional food to *end* starvation in food-deficit areas around the globe. Based on our corn yield data, the SCCS has the demonstrated potential to double world corn production. For example, in Malawi, Africa, it is relatively easy for the 2.5-ha farmer in food-deficit areas to deploy a suitable site-specific SCCS for Malawi. In the process of determining the optimum crop mix, a potentially productive tuber crop like sweet potatoes and/or grains and stubble legumes may be identified that suppress weed growth in various solar corridor row configurations and cultivar choices. We submit that expertise from Lilongwe University of Agriculture and Natural Resources (LUANAR) could collaboratively develop the solar corridor in Malawi to the extent that no one in Malawi need to go hungry.

[1] Biodiesel from the corn oil, biogas from the ruminant waste stream, more ethanol from the cellulosic kernel components, and more biogas from the ethanol plant waste stream.

Alternatively, corn-based foodstuffs could be provided to global populations facing hunger through use of state-of-the-art, highly developed crop production environments in the US Corn Belt using the SCCS that could potentially double the effective spatial unit yield of the world's most productive crop—corn. The subsequently enabled highly developed agricultural paradigm shift that concentrates the nutrient-dense food components of the corn kernel (*all* of its protein, vitamins, and minerals) into one-third of its original weight, requires less transportation fuel to get it to the protein-deprived hungry people around the world.

With an appropriate connection of the dots, an intuitive food and fuel paradigm shift can indirectly incentivize the sea change reduction of childhood obesity in the United States while preventing starvation around the world. The calories thereby removed from the human food system by this paradigm shift is no longer available to feed obesity and instead displaces British thermal unit (BTUs) of fossil fuels that source their carbon from CO_2 that cycled when dinosaurs roamed earth and exhaust their carbon combustion components into the current CO_2 while concentrating the total nutrient-dense protein, vitamins, and minerals into one-third of the volume for easier shipping to the protein deprived around the world. This synergistic approach to food and fuel systems produces more food beginning with initial deployment in this generation and reduces the projected havoc foisted on future generations by mitigating the devastation imposed on the current CO_2 and ultimate depletion of feasibly extractable finite and nonrenewable carbon resources. The fact that we have to resort to fracking and tar sands for fossil fuel recovery is an indication that we have passed the tipping point of peak oil, that is, the easily extracted fossil fuel deposits have been depleted.

While the SCCS has shown the ability to double the yield of earth's most productive crop—corn (maize)—it also has the obvious potential to simultaneously increase the total production of earth's leading source of vegetable protein—wheat. Based on the above, coupled with the ease of deployment in food-deficit areas of the underdeveloped agricultural areas of the world and with the enabling of a newly incentivized elevation of hope and the "will to succeed," our vision sees an eventual end to the following:

1. The terminal misery of human starvation, 36 million people per year (World Facts, http://www.theworldcounts.com/counters/global_hunger_statistics/how_many_people_die_from_hunger_each_year#topfacts).

2. Our current addiction to oil.
3. The current abuse of our current C cycle.
4. The current loss of our best (top) soil to erosion.
5. The current destructive forces imposed on the traditional family farm and small rural towns in the highly developed monocrop agricultural areas of the United States, and offers:

a. A unique opportunity for enhancing soil health, air and water quality for future generations, and the unleashing of earth's long-term potential bioproductivity.
b. To end starvation—the most critical of human needs. If our nearly inconceivable (2X) published yields and the third-party US Patent validated and *Agronomy Journal* published independent peer review replicated seminal results would be achieved on one-half of the world's corn acres (hectares) on which the world's most productive crop (Goldschein, 2011) is grown, at least another 400,000,000 tons of corn would be produced.[2] Furthermore the world's most important vegetable protein—wheat (Goldschein, 2011)—or other winter grains and high-protein legumes grown in the SCCS per Fig. 2.8 would be increased. Soybean yields would increase wherever they are the main crop in the SCCS. The basics of plant growth indicate that nearly all crops, when integrated into the SCCS as the main crop, would increase their yield, consistent with, but at varying magnitudes, the 2X main crop (corn) examples reported previously. Local vegetable production (and local consumption) would increase where CSA vegetable crops are floor crops. In many countries where food production fails to keep up with human needs, the solar corridor concept offers a logical fit to mitigate the most critical of human needs—food where it is critically needed for human life. Further, a robust biofuels production paradigm which effectively concentrates *all* of the protein, vitamins, and minerals of the corn kernel into one-third the volume, whereas the high-nutrient density food stock can be more easily shipped to parts of the world, where food deficits are most

[2] See details on yield performance of corn and floor crops on various soils in the US Corn Belt and impacts of the SCCS on soil biology and carbon cycling in Chapter 2, Variety selection for the solar corridor crop system (SCCS) and Chapter 4, Soil health benefits of the solar corridor crop system, and also in Deichman et al. (2016) and Kremer and Deichman (2014).

critical, to further mitigate the most critical of human needs, that is, where people are starving. All USDA-designated "food deserts" could be nourished wherever the SCCS and CSA production is successfully implemented.

Biofuels and food production

Does biofuel production affect food production and supply? Although two commonly proclaimed and easily accepted perceptions—that ethanol production takes food from the hungry and land from the rain forest—are still held by some, we'll attempt to set the record straight by simply stating: "A well-designed SCCS-based corn ethanol production paradigm enables the production of *more* high-energy, nutrient-dense protein nutrition for the malnourished, hungry, and starving" and "requires no acres to be taken from the rain forest."

In the event that the above ethanol brief statement fails in its mission, the further explanation listed below should help to clarify the aforementioned brief statement.

1. Extractable calories are the only corn-derived food components of ethanol. *All* of the high-energy, nutrient-dense protein, vitamins, and minerals remain after the ethanol has been processed and removed.
2. The remaining nutrient-dense high-energy protein, vitamins, and minerals occupy one-third of the original weight of the corn, itself, requiring less transportation fuel to get the protein to where it is critically needed for human nutrition.
3. Current corn production processed into meat, dairy, and eggs in the United States need not be taken for ethanol. It depends on the specific circumstances and choices we make but, in any event, does not require any extra corn acres, including any from the rain forest, to produce 60–90 billion annual gallons of anhydrous equivalent alcohol fuels in the United States (Deichman et al., 2016).
4. Where calories are surplus and lead to obesity instead of addressing malnutrition, hunger, or starvation, the corn ethanol paradigm should incentivize enhanced human nutrition. The subsequently enabled mitigation of childhood obesity and subsequent life of obesity should become a goal that is mutually beneficial with the goals of those who are short of protein.
5. The gargantuan potential for a growing market demand for automotive fuel energy, octane, and oxygenated fuels incentivizes the

production of more corn. And as the enhanced nutrient-dense high-energy protein, vitamins, and minerals that enable higher quality protein for human consumption becomes more available, starvation is mitigated as a result of an appropriate 360 degree corn ethanol production paradigm. If hydrogen fuel cells edge out the internal combustion engine and become the transportation fuel of choice, perhaps we can transition our crop-based feedstock into a hydrogen source for fuel cells and continue our food and fuel production roles. Accordingly, the heart of the US Corn Belt could potentially* replace all automotive gasoline used therein with E-85 without adversely taking away any rain forest acres, while enhancing the diets of protein-deficient humans and reducing obesity.

*depends on the number of acres, stakeholders and consumers that commit to the new 360 degree paradigm shift.

Soil health and environmental quality

We foresee that adaptation of the SCCS in many farming systems will have positive impacts on soil and environmental quality. However, as indicated for the many benefits of the SCCS our vision is based on the limited research conducted to date (Kremer and Deichman, 2014) on a restricted number of soils in US Corn Belt. As use of the SCCS expands globally, the multiple benefits of the system will be demonstrated in other geographic regions with contrasting soils, climate, and cultural practices.

Soil health assessment is used to monitor agricultural management systems for impacts on soil productivity that centers on the chemical, physical, and biological processes of soils and effects on environmental quality and ecosystem services. Some agricultural management practices negatively affect soil health, while other practices maintain or improve soil biological health (Lehman et al., 2015b). Thus management practices that assure healthy soils must be viewed both from a production and environmental perspective (Stirling et al., 2016). In a practical sense, soil and crop management practices must: (1) maintain soil carbon, (2) control erosion, (3) maintain soil structure, (4) maintain soil fertility, (5) increase nutrient cycling efficiency, (6) reduce export of nutrients and thus the need for increased inputs, and (7) reduce pesticide input requirements (Lehman et al., 2015a). We believe that the SCCS can fulfill all these attributes based on our seminal studies showing increases in critical soil C fractions due to inherent increased photosynthesis by the main crop; soil surface covered with residues from both main and floor crops; increased biological nutrient cycling; and potential decreased pesticide inputs due to vigorously growing floor crops (Kremer and Deichman, 2014, 2016; Kremer et al., 2015).

The SCCS potentially increases soil microbial biodiversity both from a structural and functional standpoint, based on initial studies conducted at one Missouri site (Kremer et al., 2017). This will help address the high priority needs for improving soil health and reversing soil degradation by identifying and developing the SCCS as one of the management systems for restoring, enhancing, and maintaining functional abilities of a diverse soil biological community (Lehman et al., 2015a).

The potential for management to influence soil biological communities to strengthen stability of biological processes, overall soil health, and ecosystem services will become a reality with deployment of the SCCS. Furthermore, knowledge concerning soil microbial consortia responsible for weed (Kremer and Li, 2003) and pathogen suppressive soils and the ecology of plant growth promoting rhizobacteria (Lehman et al., 2015a) will be gained effectively using the two-crop system provided by the SCCS, which provides contrasting soil and rhizosphere microbial niches on the same soil in the same field. In a relatively few cases, specialized organisms such as the arbuscular mycorrhizal fungi (AMF) have been used to demonstrate the positive effects of conservation agricultural practices like mixed-species cover cropping on soil health (Lehman et al., 2012). Other plant species also promote AMF populations and biological activity including barley, a winter-hardy species tolerant to drought, when used as a cover crop with intercropping systems (Hontoria et al., 2019) and is adaptable as a floor crop in the SCCS.

The following agricultural practices can all be included in the SCCS to modify the whole soil biological community (biomass, numbers, diversity, activity) in a positive manner: no till or conservation tillage, cover cropping, elimination of fallow, incorporation of perennial crops, retention of crop residues, diverse crop rotations, use of organic amendments, integrated pest management practices (Lehman et al., 2015a), and integration of livestock (Stirling et al., 2016).

The most important soil quality indicator for nearly all soils throughout the world is soil organic matter (SOM; Lehman et al., 2015a). Effective management involves balancing two ecological processes: mineralization of carbon (C) and nitrogen (N) in SOM for short-term crop uptake, and sequestering C and N in SOM pools for long-term maintenance of soil quality, including structure and fertility. Agricultural land management options recommended to increase SOM and improve soil quality nearly always include some reduction in tillage intensity and implementation of integrated, multifunctional cropping rotations that include forage legumes and/or small grains. Soil management practices that increase SOM and enhance soil health create expanded habitat and greater niche diversity for soil biological communities. The inputs of organic matter from plant residues and exudates provide carbon and energy sources for soil organisms. For example, no-till farming ("no-tillage") is a management practice that could readily be adapted in the SCCS and could supplement SOM buildup and increases in soil water-holding capacity (Veum et al., 2015; Wander and Bollero, 1999). We have demonstrated over a 5-year testing period the SCCS under minimum tillage indeed provides additional C to the soil microbial community and contributes to SOM buildup, and expect this process to be functional in other soils in future studies conducted in various regions (Kremer, 2019).

We also suggest that SCCS can have an impact on restoration of marginal soils degraded due to intensive cultivation. Our limited studies demonstrate that the SCCS as a sustainable farming practice can maintain and/or improve soil productivity and, over the long term, produce adequate yields without degrading soil and water resources. Once again, the attributes of a well-functioning soil presented above supports the premise that to restore degraded soils, the first step must be to enhance and maintain soil biological properties and processes, which appears to be an inherent property of the SCCS (Kremer and Deichman, 2014, 2016).

From an environmental standpoint, we envision that the SCCS will be considered to be an effective sustainable agricultural management practice. It is similar to those proposed or established that strive to integrate complexity into the management approach that includes cover crops, filter strips, and noncrop landscapes such as grasslands and forest areas that provide vital habitats for beneficial organisms and serve as nutrient sinks to capture soluble nutrients and trap contaminants before these impact aquatic ecosystems (Altieri, 1999). The positive linkages between biodiversity and provision of ecosystem services in terrestrial ecosystems that we expect to be operational in the SCCS include erosion control, regulation of nutrient cycling, soil structure maintenance, bioremediation of synthetic chemicals, mitigation of greenhouse gas emissions, enhance water quality, moderate SOM decomposition, and pest/pathogen control, as demonstrated in ecosystems exhibiting optimum soil health (Lehman et al., 2015a).

Conclusions

A summary of key benefits that we envision will be gained from an established SCCS:

- Sustainable cropping system with profit potential for beginning and advanced farmers.
- Sustainable cropping system transferrable to various global regions where corn is grown.
- Potential key to recovery of declining rural communities in the United States.
- Contribution to solving hunger problem in food-deficit regions around the globe.
- Advanced soil health improvement.
- Improved soil biodiversity—both microbial types and function.
- Continuous living roots in at least half of field to maintain living microbial community.
- Improved soil conservation, reduced erosion.
- Soil surface cover through crop residues, cover crop vegetation.
- Livestock integration—organic matter processing.
- Added value:
  - Grazing improved feed ration of combined primary and floor crops.
  - Seed harvest from floor crop.
  - Grain for livestock harvested from cereal, pulse floor crops.
  - Potential biofuel crops as main crop in conjunction with alternative floor crops.
  - Forage harvest or grazing from floor crop.
  - Cereal grain straw for bedding or market.
  - Marketable crops from corridor through farmers markets, CSA, etc., to support economy of local rural community.
  - Food available in subsistence, sustainable/ecological, or organic systems.
- Reduced or conservation tillage accommodated for improved soil structure, biological soil habitat, reduced erosion.
- Potential reduced inputs of inorganic fertilizers, herbicides, other pesticides.
- Water conservation (decrease irrigation needs—partly due to increased SOM buildup).
- Remediate pesticide residues in soils.

- Promote beneficial insects (pollinators, seed predators, pest predators).
- Weed and pest management (natural or ecologically based).
- Mitigating effects on global warming and subsequent climate change.

References

Altieri, M.A., 1999. The ecological role of biodiversity in agroecosystems. Agric. Ecosyst. Environ. 74, 19–31.

Deichman, C.L., 2019. Variety selection for the solar corridor crop system (SCCS). In: Deichman, C.L., Kremer, R.J. (Eds.), The Solar Corridor Crop System. Elsevier Academic Press, San Diego, CA, Chapter 2.

Deichman, C.L., Kremer, R.J., Lupo, A., 2016. Over 1000 gallons of liquid biofuels per acre from four different corn hybrids on a Guilford fine sandy loam. In: Proceedings, International Annual Meetings, ASA-CSSA-SSSA, Phoenix, AZ.

Goldschein, E., 2011. The 10 most important crops in the world. Business Insider, September 20, 2011, <https://www.businessinsider.com/10-crops-that-feed-the-world-2011-9>

Hontoria, C., García-González, I., Quemada, M., Roldán, A., Alguacil, M.M., 2019. The cover crop determines the AMF community composition in soil and in roots of maize after a ten-year continuous crop rotation. Sci. Total Environ. 660, 913–922.

Johnson, J.M.F., Jin, V.L., Colnenne-David, C., Stewart, C.E., Jantalia, C.P., Xiong, Z., 2017. Row-crop production practices effects on greenhouse gas emissions. In: Al-Kaisi, M.M., Lowery, B. (Eds.), Soil Health and Intensification of Agroecosystems. Elsevier Academic Press, San Diego, CA, pp. 257–275.

Kremer, R.J., 2019. Soil health benefits of the solar corridor crop system. In: Deichman, C.L., Kremer, R.J. (Eds.), The Solar Corridor Crop System. Elsevier Academic Press, San Diego, CA, Chapter 4.

Kremer, R.J., Li, J., 2003. Developing weed suppressive soils through improved soil quality management. Soil Till. Res. 72, 193–202.

Kremer, R.J., Deichman, C.L., 2014. Soil quality and the solar corridor crop system. Agron. J. 106, 1853–1858.

Kremer, R.J., Deichman, C.L., 2016. The solar corridor: a new paradigm for sustainable crop production. Adv. Plants Agric. Res. 4 (3), 00136. Available from: https://doi.org/10.15406/apar.2016.04.00136.

Kremer, R.J., Deichman, C.L., Reinbott, T.M., Veum, K.S., 2015. Soil biology is enhanced through synergistic relationships in the solar corridor production system. In: Agronomy Abstracts. ASA-CSSA-SSSA Annual Meetings, Minneapolis, MN.

Kremer, R.J., Deichman, C.L., Reinbott, T.M., Veum, K.S., 2017. The solar corridor cropping system for efficient grain production and improved soil quality on a claypan soil. In: Agronomy Abstract ASA-CSSA-SSSA Annual Meetings, Madison, WI.

Lehman, R.M., Taheri, W.I., Osborne, S.L., Buyer, J.S., Douds Jr., D.D., 2012. Fall cover cropping can increase arbuscular mycorrhizae in soils supporting intensive agricultural production. Appl. Soil Ecol. 61, 300–304.

Lehman, R.M., Cambardella, C.A., Stott, D.E., Acosta-Martinez, V., Manter, D.K., Buyer, J.S., et al., 2015a. Understanding and enhancing soil biological health: the solution for reversing soil degradation. Sustainability 7, 988–1027.

Lehman, R.M., Acosta-Martinez, V., Buyer, J.S., Cambardella, C.A., Collins, H.P., Ducey, T.F., et al., 2015b. Soil biology for resilient, healthy soil. J. Soil Water Conserv. 70, 12A–18A.

Mullally, S., 2011. Planting in 60″ rows. Sunflower 37 (6), 1–5.

Nelson, K.A., 2014. Corn yield response to the solar corridor in upstate Missouri. Agron. J. 106, 1847–1852.
Nielsen, R.L., 2017. Historical Corn Grain Yields for the U.S. http://www.kingcorn.org/news/timeless/YieldTrends.html.
Smith, D., 2000. The solar corridor: patented 60″ rows provide more sunlight, spike yields 40%. Farm J. 124 (9), special feature page 1.
Stirling, G., Hayden, H., Pattison, T., Stirling, M., 2016. Soil Health, Soil Biology, Soilborne Diseases and Sustainable Agriculture. CSIRO Publishing, Clayton South.
Veum, K.S., Kremer, R.J., Sudduth, K.A., Kitchen, N.R., Lerch, R.N., Baffaut, C., et al., 2015. Conservation effects of soil quality indicators in the Missouri Salt River Basin. J. Soil Water Conserv. 70, 232–246.
Wander, M.M., Bollero, G.A., 1999. Soil quality assessment of tillage impacts in Illinois. Soil Sci. Soc. Am. J. 63, 961–971.

Index

Note: Page numbers followed by "*f*" and "*t*" refer to figures and tables, respectively.

J

L

M

N

S

Y

Kolibri 360 GmbH
Ferdinand-Jühlke-Straße 7
99095 Erfurt, Deutschland
produktsicherheit@kolibri360.de